Astronomy for All Ages

ASTRONOMY
FOR ALL AGES

Discovering the Universe Through Activities for Children and Adults

by

Philip Harrington & Edward Pascuzzi

OLD SAYBROOK, CONNECTICUT

Illustrations and photographs are by the authors unless otherwise indicated.
Text design by Nancy Freeborn

Library of Congress Cataloging-in-Publication Data

Harrington, Philip S.
Astronomy for all ages : discovering the universe through activities for children and adults / by Philip Harrington & Edward Pascuzzi.—1st ed.
p. cm.
Includes bibliographical references and index.
ISBN 1-56440-388-2
1. Astronomy—Amateurs' manuals. 2. Astronomy—Observers' manuals—Juvenile literature. 3. Astronomy—Study and teaching—Activity programs—Juvenile literature. [1. Astronomy—Observers' manuals. 2. Astronomy projects. 3. Science projects.]
I. Pascuzzi, Edward. II. Title
QB63.H317 1994
520—dc20 93-46751
CIP
AC

Manufactured in the United States of America
First Edition/Third Printing

To my wife, Wendy, and daughter, Helen, for their eternal love and encouragement

—PHILIP HARRINGTON

To my family and friends, for all of their continued support and encouragement

—EDWARD PASCUZZI

Contents

Preface

Why did not somebody teach me the constellations and make me at home in the starry heavens which I don't half know to this day.

THOMAS CARLYLE
Nineteenth-century Scottish historian and essayist

The words of Thomas Carlyle echo a sentiment felt by many people to this day. We hope this book will serve as an invitation to inspire you to strike up a friendship with the universe.

Astronomy is a participation "sport" that must be "played" to be fully appreciated. With that in mind, we have come up with a series of get-acquainted activities for stargazers young and old. These activities range in difficulty from simple enough for elementary-school students to complex enough to challenge the inquisitive high-school mind. Some require special tools and equipment; others need only this book, a clear sky, and a sense of curiosity.

The activities are designed to be performed by adults and children together. You will find that the natural curiosity of the children's minds will spark the imagination of everyone involved. The universe is, after all, a place of wonder.

We have rated each activity by its level of difficulty. Activities that are listed as "Intermediate" are suitable for ages eleven through fourteen. "Advanced" activities are appropriate for ages fifteen and up. Activities that are marked with the word "All" are suitable for an elementary level (ages seven through ten) as well as for all students of astronomy from the age of eleven on up.

Start with the basic naked-eye activities in the beginning of the book. You can learn to recognize the night sky as you do your daytime world. It has been said that most people go through life looking either straight ahead or down toward the ground. By doing so, they're missing half of their world. What a shame, especially when the other half—that is, the half over their heads—has so much to offer.

Once you're familiar with the naked-eye night sky, graduate to those activities that require more in-depth observations of the heavens. Some of these latter activities require a telescope or binoculars. If you don't own a telescope or binoculars, perhaps consider buying the needed equipment (see Activity 45). Or, if you are so inclined, consider building the telescope we detail in Activity 46. Making a telescope is a great family activity.

After you do these activities, we invite you to tell us your results and reactions. Please send your comments to us in care of the publisher, The Globe Pequot Press, P.O. Box 833, Old Saybrook, Connecticut 06475.

Here's wishing you clear skies and bright stars. Welcome to the universe.

Philip Harrington
Edward Pascuzzi

Acknowledgments

The book you hold is not simply the product of two authors, but instead a collaboration between the authors and the hundreds of children and adults we have known and taught over the past two decades. We are especially thankful to the students who attended our classes at Vanderbilt Planetarium, in Centerport, New York. Though they may not have known it at the time, they field-tested many early versions of these activities. We are also grateful to Mark Levine, director of the planetarium, for his consideration and helpfulness.

Our sincere thanks go to Sam Storch, director of the Hubble Planetarium at Edward R. Murrow High School in Brooklyn, New York, and a lifelong astronomer. He carefully read over our manuscript and suggested many important changes. In addition, Clifford Swartz, professor of physics at the State University of New York at Stony Brook, reviewed and checked several of the activities.

Many of the photographs in this book were taken not by professional astronomers in observatories, but by amateur astronomers. For these efforts, we thank John E. Bortle, Susan and Alan French, Richard Hill, Brian Kennedy, Charles Layman, Allan E. Morton, Gary Oleski, Greg Terrance, George Viscome, Frank Zullo, and especially Dennis Milon.

We would like to thank our editor at The Globe Pequot Press, Laura Strom, for her guidance, enthusiasm, and suggestions.

And finally, we give love and thanks to our families for their patience and support throughout. They were always there, even on the cloudiest nights.

SECTION I

The Naked-Eye Sky

1 Finding Your Way in the Sky

Level: All

Objective: To learn the four compass directions

Materials: None

BACKGROUND

Before we begin to delve into the universe, it is important to become familiar with our own little corner. The next several activities will introduce you to the night sky, but in order to understand and use this information, it is important to learn a little about the sky itself.

Before you read any further in this book, go outside and look up. Take a look at the daytime sky. What do you notice about it? What is its shape?

To many people, the sky appears like an upside-down bowl hanging over their heads. Using this analogy, the edge of the bowl (that is, the place where the sky meets the ground) is called the **horizon.** The horizon extends completely around the sky.

Everything in the sky that is beyond our atmosphere appears to cross the sky very slowly from east to west. This is caused by Earth' s rotation. Earth spins on its **axis** once every twenty-four hours. Earth's axis is an imaginary line that passes from the North Pole, through the center of the Earth, to the South Pole. As Earth rotates, the stars, sun, moon, and planets appear to move by, rising in the east and setting in the west.

ACTIVITY

This and the next activity involve finding our way around the sky. Let's begin by locating the four directions, or compass points: north, south, east, and west. Go outside on the next clear evening just as the sun is setting, as shown in Figure 1-1. Notice the spot where the sun is dropping below the horizon. (Warning: Do NOT look at the sun directly.) Remember, the sun (and just about everything else in the sky beyond Earth's atmosphere) appears to set in the west. Make a mental note of where the setting sun is from your observing spot. That direction is west.

With west established, the other three directions will come easily. If you look at any map, either of Earth or of the sky, you will notice that the direction east is opposite west. Consequently, as you are facing the setting sun in the western sky, the east point on the

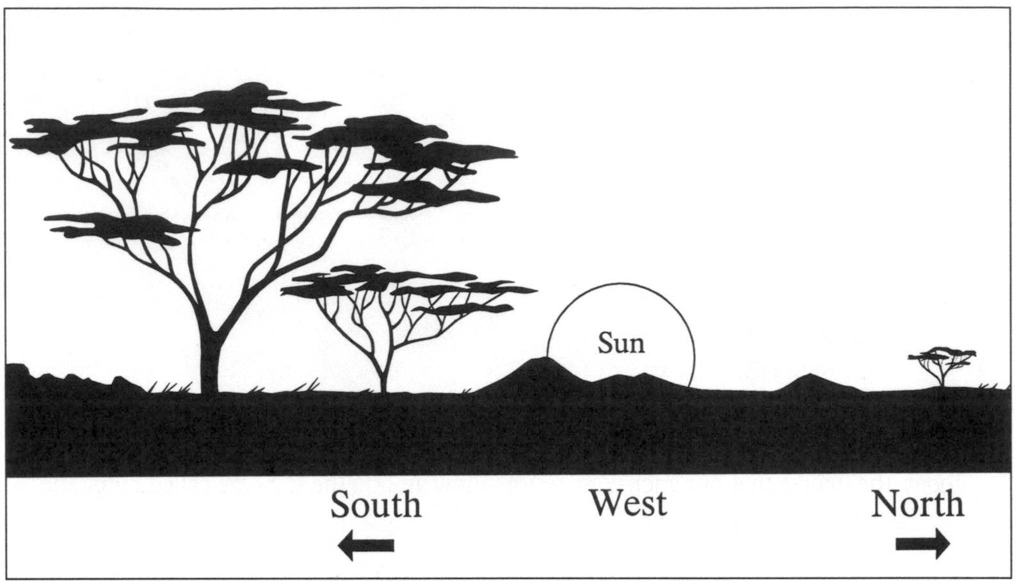

Figure 1-1. The four compass points can easily be determined from the setting sun. Looking toward the sunset point, north is to the right, south to the left, east directly behind, and west directly in front.

horizon is directly behind you. Turn completely around so that the setting sun is toward your back. You will then be facing east.

Turn back around to the setting sun in the west. To find north, extend your right arm straight out from your side. Your hand will be aimed to the north. Next, stretch out your left arm. That hand will be pointing toward the south.

Finally, look straight up. Astronomers call the point directly overhead the **zenith.**

That's all there is to it. Now that you know a few basic terms, and with the sun setting, read the next activity on how to use star charts to "read" the night sky. Then, selecting the appropriate seasonal activity from numbers 4 through 7, go out after it gets dark tonight (if it's still clear, that is) and begin to identify some of the stars and star patterns. With this introduction and the maps provided, you should have little trouble finding your way around the sky.

2 Using a Star Map to Read the Sky

Level: All

Objectives:

- To learn how to read and use a star map
- To make an astronomer's flashlight

Materials:

- An all-sky star map (from Activities 4 through 7)
- A small flashlight
- Some red filter material, such as cellophane
- A rubber band

BACKGROUND

Think of this activity as a warm-up for Activities 3 through 7. Those five activities will take you on a year-long journey to foreign lands, where you will meet exciting people and encounter creatures, some delightful, some ferocious. You will see sights that most people have never seen, yet you will journey to these places without ever leaving your backyard. You will journey to the night sky!

Thousands of years before the age of movies, radio, television, and VCRs, our ancestors gathered at night and told exciting myths and legends. These elaborate stories occurred in exotic places and involved some amazing creatures and beings. Since the printing press was still more than a millennium away, these tales were passed from one generation to the next by word of mouth. Storytellers would illustrate their narratives using figures drawn among the stars.

Today, the figures that our ancient ancestors drew among the stars are called **constellations.** There are eighty-eight constellations in all. Nobody knows who devised the first constellation, but most come to us from ancient Greece, Rome, Egypt, Persia, and Babylonia. Other cultures, such as Native American and those of the Orient, also made up constellations, but most of their patterns were different from the Western ones we recognize today.

Different constellations are seen at different times of the year. Those seen on spring evenings are called the spring constellations, and so on. Some constellations seen in the northern part of the sky never set when viewed from the mid-northern hemisphere vantage point of the continental United States. These are referred to as **circumpolar** constellations. While shown on all four seasonal maps in this book (Activities 4 through 7), these constellations are discussed separately in Activity 3. Still others never rise from middle latitudes, as they are too far south to peek above the southern horizon. To view these constellations, a resident of the northern hemisphere would have to travel to lands south of the equator.

To eliminate the clutter and confusion often associated with maps of the stars, only the brighter, more easily recognized constellations are plotted on the star charts in Activities 3 through 7. The constellation figures are marked by dashed lines between their stars; their names are printed in capital letters. Use these maps and the accompanying text to guide your way across the sky. Throughout, many of the ancient tales are recounted, sprinkled with a dash of modern astronomy.

ACTIVITY

Which star map should be used at which time of year? Take a look along the bottom of each seasonal chart (Activities 4 through 7) at the time schedule. The spring map, for instance, shows the sky as it will appear in early April at midnight. The same stars will be seen at 10 P.M. in early May and at 7 P.M. in late June.

Keep in mind that these maps are drawn for the way the sky will look during evening hours. In the wee hours of early morning, the stars and constellations of the next season would replace those shown on the current seasonal map. If, for instance, you were to go out at, say, 3 A.M. in late April, you would not see the spring constellations, but rather the groups shown on the summer chart.

Notice how some star symbols appear larger on the maps than others. This is a convention used on all star charts to depict star brightness, or **magnitude** (see Activity 9). The brighter the star, the larger its symbol. Also realize that despite their appearance on our star maps, real stars do not have five points—they are actually spheres of hot gases. Many of the brightest have names assigned to them. Each name is shown in small print adjacent to the star.

Before departing on our star trek, it is important to understand how to use a star map. To begin, look at any one of the four seasonal charts found in Activities 4 through 7. Notice the trees and buildings along the circular border of the map. This represents the complete **horizon,** the line where the sky meets the ground. The center of the chart marks the sky's overhead point (called the **zenith**), while around the horizon are the four compass points—north, south, east, and west.

The maps are drawn for approximately 40° north latitude, which is the latitude for an east-west line passing from about New York City to somewhat north of San Francisco. If you live south of this line, some stars will be seen above your southern horizon that are not shown on these maps. If you live to the north, then the stars in the south will be shifted closer to the horizon. In both cases, corresponding changes take place along the northern horizon as well. However, this slight difference will not affect the maps' usefulness within the continental United States.

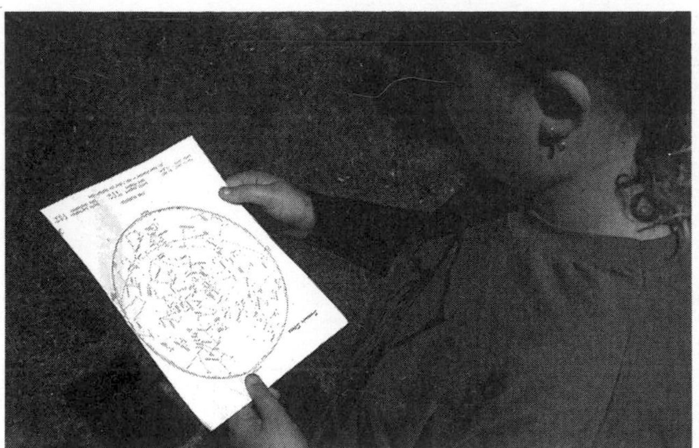

a

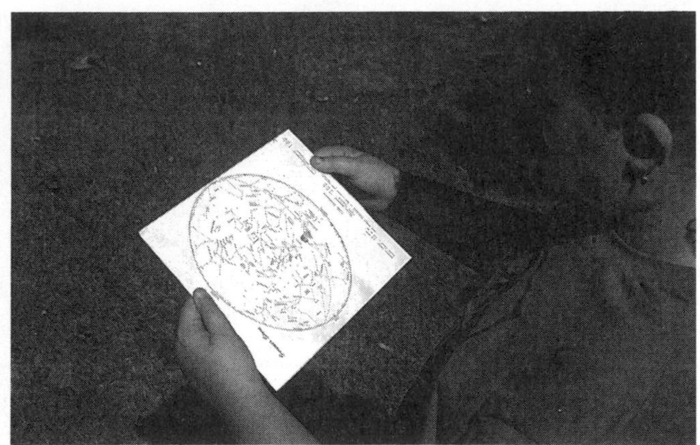

b

Figure 2-1. Helen Harrington demonstrating how to hold a star chart correctly to view (a) the north sky, (b) the east sky. Note the angle of the chart itself.

The best way to use the maps outdoors is to rotate the book so that the name of the direction you are facing is angled toward the ground, as show in Figure 2-1. If you are facing south, hold the book upright, so that the word "south" (and the time schedule) are nearest to your body. But if you are facing north, turn the page upside down, so that the word "north" is now nearest to you. Rotate the page a quarter-turn either left or right, as required, for east and west. (Remember, when finding directions, that the sun rises in the east, sets in the west, and is in the south at lunchtime.)

The tilt of the constellation and star names will also help aim the chart in the right direction. By rotating the book around until a name appears horizontal, the chart will be oriented correctly for that star or constellation.

You'll need a flashlight to read the charts outdoors at night. While any flashlight will do the job, the beam must first be covered with red transparent material. An unfiltered (white) light, even a dim one, can become blindingly bright if you have been in the dark for a while. Just think of the last time you woke up and decided to raid the refrigerator for a midnight snack. It probably took a minute or more to get used to the increased brightness before you could see without squinting. Then, once the light was shut off, it took another minute or more for your eyes to become accustomed to the darkness again.

This process is called **dark adaptation.** The human eye can take up to half an hour or more to become fully dark adapted, which is why stargazers carefully guard their night vision. Experience shows that a red light affects night vision less than a white light. That's why astronomers prefer red flashlights to white lights. To make an astronomer's flashlight, cover the flashlight beam with a piece of red transparent material such as cellophane, available from most stationery, party-goods, and art-supply stores. Another good source of red filter material is automobile taillight repair tape, sold by most auto parts store. Even a cut-up red balloon placed over the end of a small penlight will work well.

If you choose red cellophane, cut a square about an inch larger than the front diameter of the flashlight. Wrap the cellophane evenly around the front of the flashlight and secure it in place with a rubber band. If you decide to use the auto lens repair tape, disassemble the flashlight and cut a piece of tape the same size as its clear plastic window. Stick the tape to the window, trimming it as required, and reassemble the flashlight. Voila—an astronomer's flashlight!

The night sky awaits your visit. If it is going to be clear tonight, why not plan an observing session to become acquainted with the stars? The following five activities detail the brighter stars and constellations (or star patterns) visible throughout the year. Take a look at each, then select the appropriate activity for the current season. After darkness falls, begin your exploration of the sky. You will find, as we have, that we live in an amazing and wondrous universe.

If you plan to show the stars and constellations to groups of people, it can be difficult to point out the objects simply by saying "look over there." In this case, we recommend using a bright, unfiltered spotlight-style flashlight. If the light is bright enough, the beam will act like a pointer. Aim the light into the sky toward the star or constellation under discussion and tell the others to follow the beam. Just be careful not to aim the unfiltered light at anyone's eyes or at your star map, or—POOF—no more night vision!

3 Locating the Constellations and Bright Stars of the Circumpolar Sky

Level: All

Objectives:

- To find the north circumpolar constellations
- To learn the associated myths and legends

Materials:

- A clear sky anytime of the year
- A red-filtered flashlight (see Activity 2)
- Circumpolar star map (Figure 3-2)

Anyone can easily see that the sun, moon, planets, and stars appear to rise in the east, arc slowly across the sky, and set in the west hours later. This apparent motion is caused not by the heavens themselves moving but by Earth rotating on its axis. As Earth revolves around the sun, our world's night side faces toward different parts of the cosmos. The result is that different stars are visible both at different times of the night and at different times of the year.

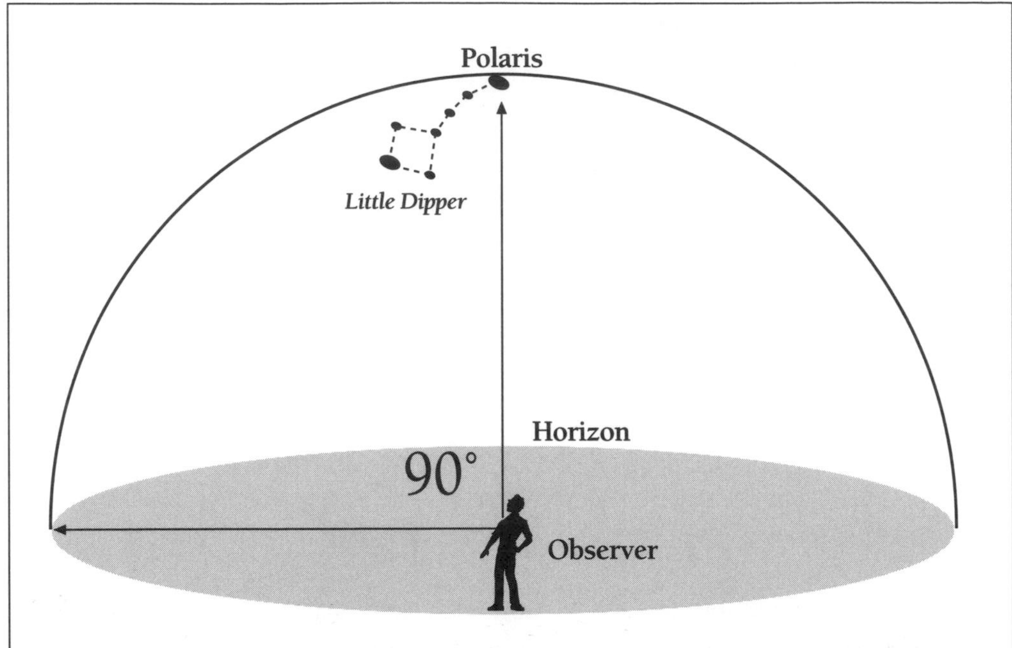

FIGURE 3-1. The Little Dipper seen from (a) the North Pole (latitude 90° north) and, (b) from latitude 45° north.

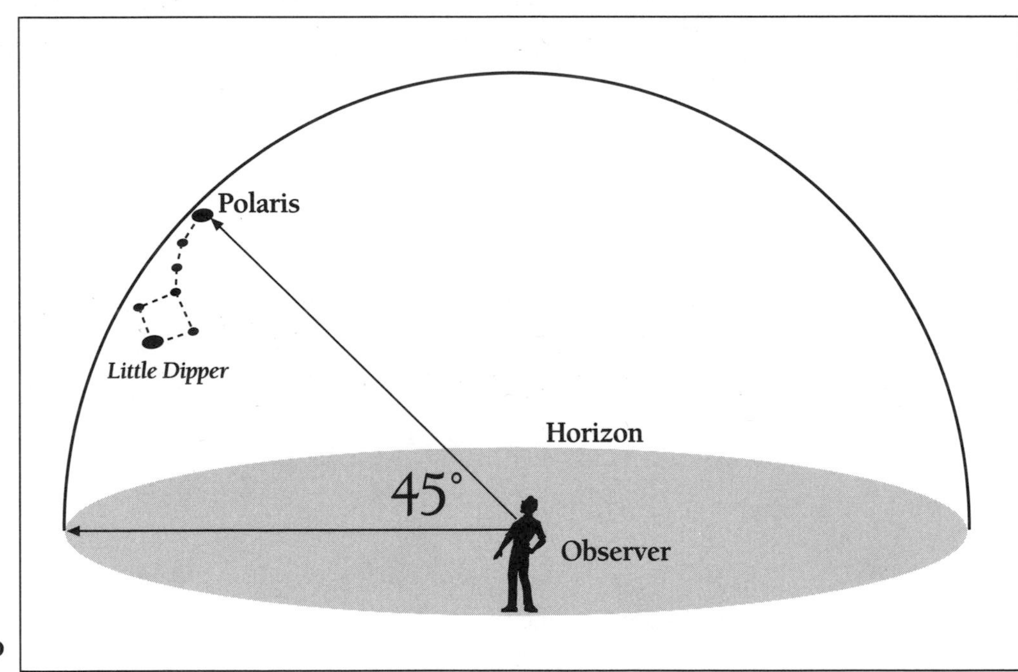

Yet not all stars rise in the east and set in the west; in fact, some constellations never set at all. If we were located at either of Earth's poles (as in Figure 3-1a), the stars would never rise nor set; instead they would move parallel to the horizon. As we travel away from Earth's poles toward the equator, the sky slowly "tilts over." Residents of the continental United States live roughly halfway between the north pole (90° north latitude) and the equator (0° latitude). From this vantage point, the sky appears angled at 45°, as in Figure 3-1b. As a result, stars that lie closer to the North Star, Polaris, will never set. Instead, they will curve around underneath, never quite touching the horizon, then rise up again in the northeast. These stars, and their associated constellations, are referred to as **circumpolar.**

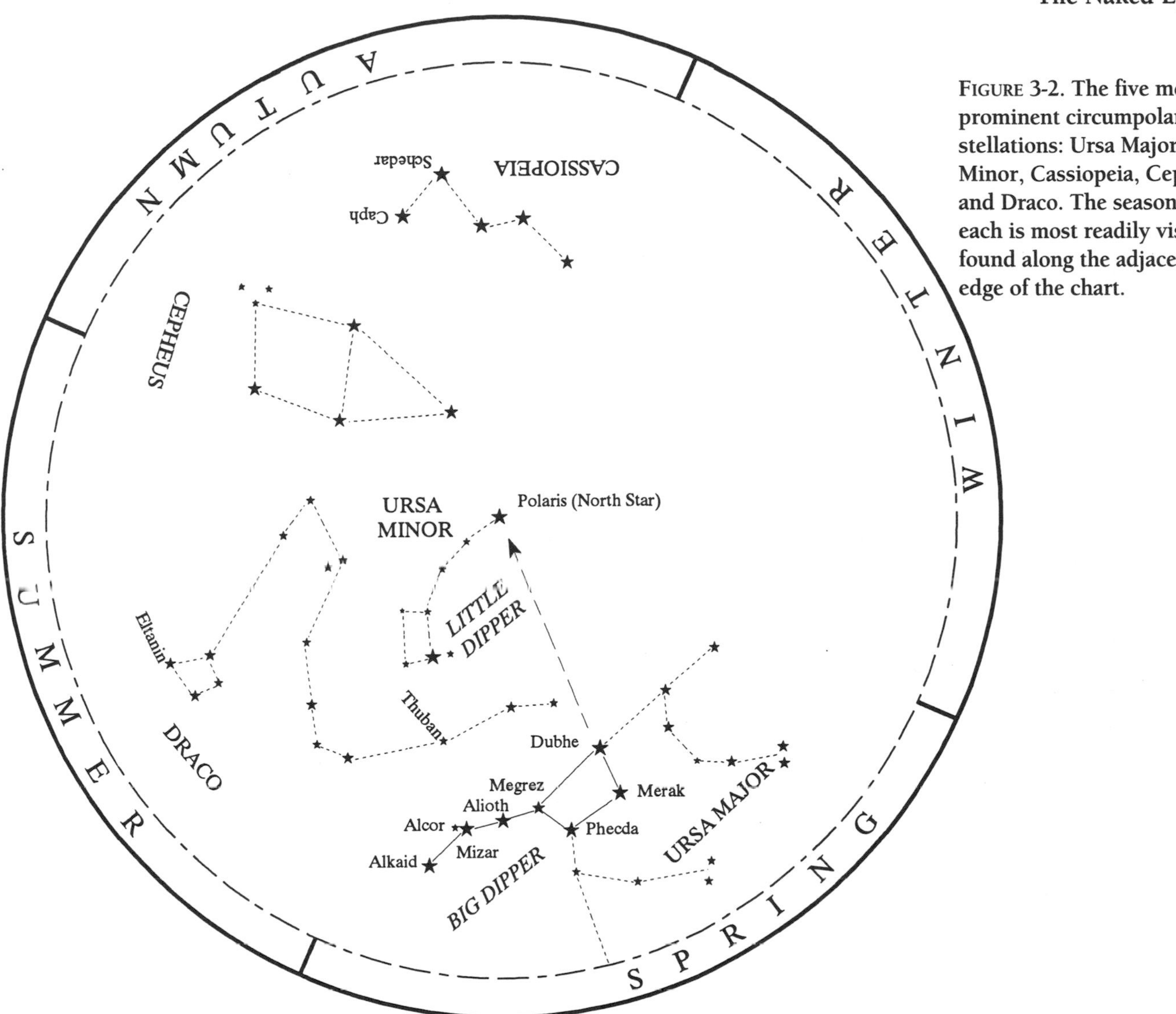

FIGURE 3-2. The five most prominent circumpolar constellations: Ursa Major, Ursa Minor, Cassiopeia, Cepheus, and Draco. The season when each is most readily visible is found along the adjacent edge of the chart.

Only five prominent constellations, depicted in Figure 3-2, populate the north circumpolar sky. (Southern hemisphere astronomers have their own south circumpolar counterparts, but none are visible from mid-northern latitudes.)

Although they remain visible all year, different circumpolar constellations are seen better during certain seasons than others. For instance, spring is the best time to see the best-known circumpolar figure of all, the **Big Dipper.** Known by every Boy Scout and Girl Scout, the Dipper is formed by seven bright stars: **Megrez, Dubhe, Merak,** and **Phecda** mark its cup; **Alioth, Mizar,** and **Alkaid** denote its crooked handle. Together they are found in the northeast on cold winter nights, overhead on spring evenings, in the northwest during summer, and scraping the northern horizon in the autumn.

The Big Dipper by itself is not an official constellation but part of a much larger star picture named **Ursa Major.** The name "Ursa Major" means "Great Bear." Can you imagine a monstrous bear in this part of the sky? Perhaps, with a little imagination. The bowl of the Dipper forms the bear's back and belly. From the star Dubhe, faint stars extending away from the bowl make up the bear's neck and head, while curving below (to the south) are five stars forming one of its front paws. Extending below (to the south) of the bowl star Phecda are other faint stars that represent its hind legs. In this picture, the three stars in the Dipper's handle usually create the bear's tail.

Many different and completely unrelated ancient cultures coincidentally pictured this group of stars as a bear, creating legends that told of its ferociousness. Perhaps the most interesting comes to us from the Iroquois Indians, who tell the story of how every spring a bear leaves its den to wreak havoc on their tribe. The Indian chiefs decided to send their three mightiest warriors out after the bear. (In this story the stars in the Dipper's handle represent the warriors rather than the bear's tail.) The warriors chase the bear across the sky throughout the summer. Finally, in the autumn, as the bear tires, one of the Indians wounds it with an arrow. Blood drips from the wound and colors the leaves of the forest in shades of red, orange, and yellow. But the wound is not fatal—and both the bear and the relentless hunters return to the sky every spring.

Take a careful look at Mizar, the middle star in the Dipper's handle. With keen eyesight, you might be able to make out a second, fainter star just above it. As shown on the map, this companion star is named **Alcor.** Together Alcor and Mizar form what astronomers call a **double star**, two stars situated very close to each other. Estimates show that as many as half of the stars seen are double stars. Most of these twosomes are true stellar pairings, called visual double stars, with each star gravitationally bound to the other. Some, however, are only chance alignments seen from our vantage point and are called optical doubles. Alcor and Mizar constitute perhaps the most famous example of an optical double star. Although they appear very close together, Alcor and Mizar are nowhere near each other in space. (Interestingly, Mizar is a visual double star as well, with a faint companion visible in most small telescopes. More about this in Activity 40.)

Polaris has not always marked the so-called north celestial pole. Due to a 26,000-year wobble of Earth's axis, the north celestial pole appears to trace a circle in the sky. Right now, the pole happens to align almost perfectly with Polaris. More than 4,500 years ago, when the ancient Egyptians built the pyramids, the pole aligned closely with the star THUBAN in the neighboring constellation Draco. In 14,000 years, the pole will be aimed toward Vega in the summer constellation Lyra. Twenty-six-thousand years from now, Earth's north pole will again be aimed toward Polaris.

With the Big Dipper such a conspicuous and easily recognizable set of stars, it can be used to help point us toward other, less apparent stars and patterns in the sky. For instance, take a look at the stars Merak and Dubhe in the bowl. These two stars are nicknamed the pointer stars because if you draw a line from Merak to Dubhe and extend it above the Dipper, the line points right at the **North Star.** Many people are under the mistaken impression that the North Star, properly known as **Polaris,** is the brightest star in the night sky. Not so. In fact, the North Star is the forty-ninth brightest star in the sky, just an average-looking star set among countless others. What makes the North Star special is that Earth's north polar axis is aimed almost directly at it. As a result, Polaris appears fixed in the night sky, while all other stars appear to rotate around it.

Polaris is at the end of the handle of the **Little Dipper.** Just as the Big Dipper is part of larger Ursa Major, so too is the Little Dipper part of the Small Bear, **Ursa Minor.** Be it a bear or a dipper, try to find all seven stars that are shown on the map. They are faint and will require dark skies and good eyes to be seen. Suburban amateur astronomers frequently use the visibility of all seven stars to gauge the clarity of the night sky. If the entire Little Dipper can be seen, the sky is exceptionally free of haze and extraneous lights.

On the opposite side of Polaris from Ursa Major lies another prominent circumpolar constellation that is most easily seen in the autumn. **Cassiopeia** the Queen is formed by five stars set in a stretched W pattern. Accompanying Queen Cassiopeia is her husband, **Cepheus the King.** The stars of Cepheus, all much more difficult to pick out than those of Cassiopeia's W, frame what looks like a simple drawing of a house, with the peak of the roof pointing roughly toward Polaris.

According to legend, Cepheus and Cassiopeia ruled over ancient Ethiopia. Cassiopeia was well-known for two things: her great beauty and her unabashed boastfulness. It was this unrelenting bragging that later caused Cassiopeia and Cepheus great trouble, but this is a story for the autumn sky in Activity 6, when other characters will also be visible among the stars.

Winding its way from Cepheus to the pointer stars of the Big Dipper is **Draco the Dragon.** Draco was one of the first constellations to be created among the stars and one of the few that actually looks like what it represents. Four faint stars create the dragon's head, while its long, thin body extends toward Cepheus, coils around Ursa Minor, and slithers between the bears. Though visible all year, Draco is most easily seen in the summer. Near the end of the body of Draco is Thuban, the polestar at the time of the Egyptian pharoahs. Can you find Thuban in the real sky? Notice how far away it is from our polestar, Polaris.

The circumpolar stars and constellations are visible throughout the year on every clear night, making them ideal for new stargazers to observe and learn.

4 Locating the Constellations and Bright Stars of Spring

Level: All

Objectives:
- To find the prominent constellations visible in the spring
- To learn the associated myths and legends

Materials:
- A clear springtime night sky
- A red-filtered flashlight (see Activity 2)
- A seasonal star map (Figure 4-1)

Spring is an exciting season both on the ground and in the sky. Here on Earth, the cold, barren days of winter are coming to an end. Life returns to the land as the season's warmth bathes newly blossomed trees and flowers. Overhead, our view of the night sky reveals several bright stars and many more fainter points of light.

Passing near the zenith are the seven stars that form the Big Dipper, part of the constellation Ursa Major (see Activity 3). Just as the stars of the Big Dipper were used in Activity 3 to find the North Star, Polaris, so too may they be used to locate some springtime stars and constellations. For instance, return to the Dipper's bowl. Use the bowl stars Megrez and Phecda to find a lion in the sky. Extend a line from Megrez to Phecda southward to a bright star. That star is **Regulus,** the brightest star in **Leo the Lion.** According to legend, Regulus marks the lion's heart. Above Regulus are five stars that combine to form the lion's head, though today, they remind us more of a sickle or a backward question mark. The body of Leo stretches out to the left (east), with the animal's hindquarters and tail represented by a triangle of three stars. The name of the triangle's easternmost star, **Denebola,** comes from the Arabic **Al Dhanab Al Asad,** "the lion's tail."

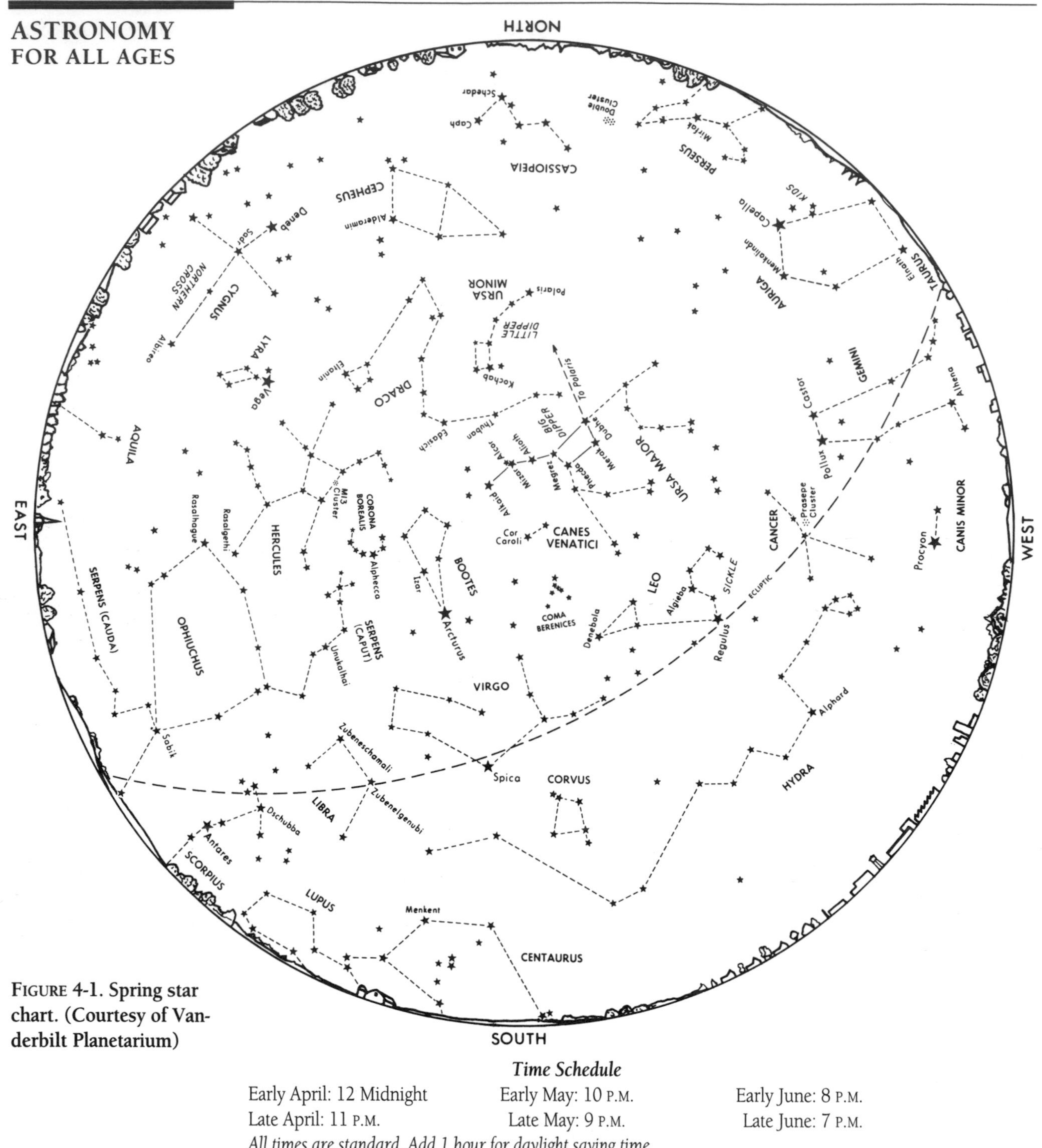

Figure 4-1. Spring star chart. (Courtesy of Vanderbilt Planetarium)

Time Schedule

Early April: 12 Midnight	Early May: 10 P.M.	Early June: 8 P.M.
Late April: 11 P.M.	Late May: 9 P.M.	Late June: 7 P.M.

All times are standard. Add 1 hour for daylight saving time.

Continue the line from the Dipper through Regulus and farther toward the southern horizon. There lies another bright star, **Alphard,** and its constellation, **Hydra.** Hydra was a horrific, many-headed serpent that inhabited the swamps of Lerna, Greece. After many failed attempts, the mighty warrior Hercules slayed Hydra, but only after he killed each head individually. The sky's Hydra has but one head, formed by a small trapezoid of faint stars to Alphard's northwest. Hydra's long but very faint body winds its way along the southern horizon. From head to tail, Hydra stretches a full third of the way around the sky, making it the longest constellation of all.

Return to Regulus momentarily, then look toward the northwestern horizon for a pair of bright stars. Those are **Castor** (on the right, or north) and **Pollux** (on the left, or south), twin stars in the constellation **Gemini.** (Gemini is a winter constellation; see Activity 7.)

Draw a line between Regulus and Pollux, pausing about halfway along. From darker suburban and rural observing sites, you just might be able to pick out the five faint stars that form the inconspicuous constellation **Cancer the Crab,** a crab sent by Hera to hinder Hercules in his battle with Hydra. If you have a keen eye, you might also notice a faint smudge centered in the body of the crab. Astronomers call this the Praesepe cluster. Though it doesn't look like much to the eye alone, that smudge is of great interest to observers with binoculars and telescopes. But why spoil the surprise? We will return to this spot in Activity 41.

Cancer, like Leo above, is one of the twelve constellations of the **zodiac,** or **ecliptic.** The ecliptic is an invisible line that circles the sky, marking the plane of our solar system. The sun, moon, and all of the planets (save for tiny Pluto) are always found somewhere close to the ecliptic. Any constellation through which the ecliptic passes belongs to the zodiac.

Turn back to the Big Dipper. Draw an imaginary line through the three stars that form its handle and continue the arc counterclockwise until you arrive at brilliant **Arcturus.** Arcturus, an orange-colored star, is the brightest star of the spring sky.

Arcturus resides in the constellation **Boötes the Herdsman.** Though our modern-day minds may have a hard time imagining a human form among the stars of Boötes, most people can picture a kite or even an ice cream cone. If you look carefully, you just might be able to spot an extra scoop of ice cream formed by a semicircle of stars to the east.

That so-called "scoop" is actually another constellation, **Corona Borealis,** the **Northern Crown.** In mythology, this jeweled crown belonged to Princess Ariadne, daughter of King Minos of Crete. When Ariadne died, her crown was placed in the sky as a memorial.

On the opposite (western) side of the herdsman, set between Bootes and Leo, is a faint mist of stars. This is the constellation **Coma Berenices,** the golden hair of the Egyptian Queen Berenice, wife of Ptolemy III. One day, as her husband rode off to do particularly fierce battle, Berenice vowed to the goddess Aphrodite that she would sacrifice her beautiful hair if her husband was permitted to return to her safely. After he did, Berenice kept her promise by cutting off her hair and presenting it to the gods. But the hair mysteriously disappeared overnight. Thinking quickly to avoid alarming the royal couple, Conon, the court astronomer, pointed to the night sky, toward our constellation of Coma Berenices. He told the king and queen that the misty appearance of that part of the sky was actually Berenice's hair and that Aphrodite had placed it among the stars for all the world to see.

Back to the Dipper we go. Retrace the arc to Arcturus, then continue to speed on to **Spica.** Take a careful look at the colors of both Spica and Arcturus and note their difference. A star's color depends on its temperature. Spica's blue-white tint tells astronomers that it is much hotter than orange Arcturus.

Spica belongs to the zodiacal constellation **Virgo the Maiden.** Virgo was recognized by many ancient cultures as either a goddess or royalty. Ancient Egypt associated her with Isis, goddess of the heavens; the Chaldeans revered her as the Queen of the Stars; in India, she was known as Kawni, mother of the god Krishna. Most often, however, she is portrayed as the Roman goddess Persephone, goddess of the harvest, with Spica marking an ear of wheat held in her left hand.

Once again, return to the Big Dipper's handle. Follow the arc to Arcturus, speed to Spica, then continue to curve toward **Corvus the Crow.** Though its stars are faint, Corvus is sur-

Arcturus holds a unique footnote in twentieth-century history. During the opening ceremonies of the 1933 "Century of Progress" world's fair in Chicago, light from Arcturus was used to trigger photoelectric cells, switching floodlights on to signal the start of the exposition. Arcturus was chosen because, at the time, it was believed to be forty light years from Earth (a little farther than the currently accepted value of thirty-seven light years). At that distance, the starlight reaching Earth in 1933 left Arcturus in 1893, the same year a similar exposition was held in Chicago.

prisingly easy to spot low in the southern sky, riding on the back of Hydra. To see a crow here, however, do not think of a trapezoidal pattern as shown on the star map. Rather, imagine a cross tilted on one side. The crow's body extends between the upper right (northwest) and lower left (southeast) stars. Its outstretched wings spread from the upper left (northeast) to the lower right (southwest) stars.

The late-spring constellation **Libra the Balance Scale** (sometimes also called the "scales of justice") is formed from another inconspicuous trapezoid of stars. Look between Spica and the southeastern horizon. Like Cancer, Leo, and Virgo, Libra is a member of the zodiac. And though it is one of the most difficult zodiacal star patterns to spot, Libra carries the distinction of being the only inanimate member of this select group of twelve constellations.

Long ago, Libra was not a separate constellation. Instead, its stars formed the claws of summertime's **Scorpius the Scorpion.** In fact, the names of Libra's two brightest stars, **Zubenelgenubi** and **Zubeneschamali,** translate as the "southern claw" and "northern claw."

With the spring come warmer nights that are ideal for stargazing. Take advantage of these more moderate temperatures to meet some unique personalities.

5 Locating the Constellations and Bright Stars of Summer

Level: All

Objectives:
- To find the prominent constellations visible in the summer
- To learn the associated myths and legends

Materials:
- A clear summer night sky
- A red-filtered flashlight (see Activity 2)
- A seasonal star map (Figure 5-1)
- A blueberry pie (optional)

Our summer sky is dominated by three brilliant stars—**Vega, Deneb,** and **Altair**—set in a large triangle. This is the **Summer Triangle;** not an "official" constellation, just an interesting pattern among the stars. Each of the three stars in the Summer Triangle belongs to a separate constellation.

Brightest of the Triangle's stars is **Vega,** found in the middle of the summer star map and high overhead during July, August, and September. Vega, a brilliant blue-white star glistening like a diamond, belongs to the constellation **Lyra the Lyre**. This is the mythical musical instrument of Orpheus, son of the sun god Apollo. Apollo taught Orpheus to play the instrument so beautifully that even savage beasts were soothed into submission. After Orpheus died, his lyre continued to play lyrical music and so was placed in the sky by the gods for all to see. (Our word "lyrical" comes from the name Lyra.) In most depictions, Vega

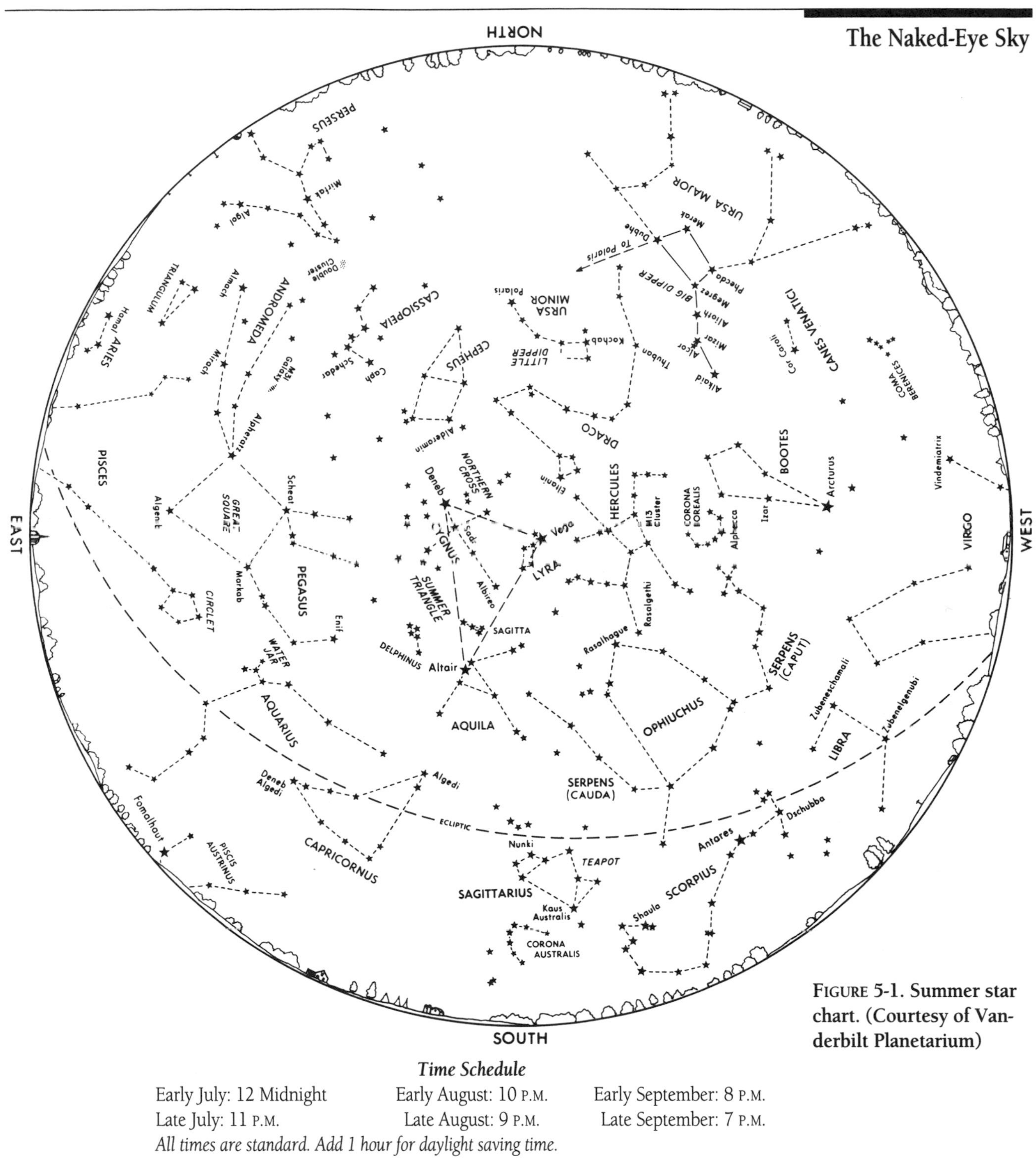

FIGURE 5-1. Summer star chart. (Courtesy of Vanderbilt Planetarium)

Time Schedule

Early July: 12 Midnight	Early August: 10 P.M.	Early September: 8 P.M.
Late July: 11 P.M.	Late August: 9 P.M.	Late September: 7 P.M.

All times are standard. Add 1 hour for daylight saving time.

represents part of the lyre's handle, while four faint stars depict the instrument's main body.

Leave the Summer Triangle temporarily by drawing an imaginary line from Vega to the springtime star Arcturus, still seen in the northwest. About a third of the way along that line are the faint stars of **Hercules the Giant.** All are visible only from darker suburban or rural skies. Four stars nicknamed the **Keystone** create his torso. Two curved lines of stars extending southward (below) form his arms, while two other star-curves stretch out to the Keystone's north to form his legs. That's right, the celestial Hercules is standing upside down, apparently on the head of Draco the Dragon.

Positioned along the western border of the Keystone is an exceedingly faint but interesting object that astronomers call the M13 cluster. Under the best sky conditions, M13 can be just made out with the naked eye as a tiny smudge. Just what is M13 and what makes it so interesting? We will return here to answer these questions in Activity 41.

Return to the Summer Triangle and locate the star Altair, its southernmost star. Altair lies within **Aquila the Eagle,** marking the beak of the bird, while fainter stars denote its outstretched wings and tail feathers. As the map implies, it might be easier to see a pterodactyl here than to imagine an eagle!

The third star in the Triangle is Deneb, in **Cygnus the Swan.** In this case the shape of a swan captured in mid-flight is quite easy to imagine. Deneb represents its tail. The swan's body and long neck extend to the star **Albireo** (itself a fascinating sight in telescopes; see Activity 40). Wings stretch outward from either side of the star **Sadr** in the swan's body. Many people also refer to this pattern as the **Northern Cross** for its likeness to a crucifix.

If you are viewing from darker suburban or rural skies, look for a hazy band of light passing through the Summer Triangle and stretching from the northeast to the south. This is the **Milky Way.** No, not a candy bar, the Milky Way is our galaxy—a pinwheel-shaped system of perhaps 400 billion stars, of which the sun is just one. Actually, all of the stars seen in the sky belong to the Milky Way. It is just that when we look along the edge of our galaxy, we are seeing many times the number of stars than when we look above or below it.

To help visualize this, think of the Milky Way galaxy as a blueberry pie. Imagine that each individual blueberry in the pie is a star and that Earth is orbiting one of those stars about two-thirds of the way out from the center. Clearly, we will see many more blueberries when we look either along the edge or toward the center of the pie than when we look up or down at it.

Follow the Milky Way down toward the southern horizon and the constellation **Sagittarius the Archer.** Sagittarius is pictured as a centaur; that is, a mythical creature with the upper body of a man and the lower body of a horse. Most artists draw Sagittarius facing toward the right (west), his bow and arrow aimed toward the heart of the Scorpion, Scorpius. Perhaps Sagittarius misfired his first arrow, for we see the tiny constellation **Sagitta** the Arrow lying just above (north of) Altair. Sagittarius may have represented a centaur to our ancestors, but to skywatchers today its outline is more reminiscent of a stove-top kettle or teapot.

As we gaze toward Sagittarius, our sights are set on the center of our Milky Way galaxy. Given dark skies and a pair of binoculars, the entire region explodes in stardust—a magnificent sight to be revisited often.

To the right (west) of the Sagittarius teapot lies the bright star **Antares,** the heart of **Scorpius the Scorpion.** The orange-red color of Antares should be immediately obvious to most stargazers. Antares is known to be a red supergiant star, nearly 2,000 times larger than our sun though only about half as hot. The curve of the scorpion's body extends toward the southern horizon and ends at the star **Shaula,** marking its poisonous stinger.

According to legend, Scorpius killed the mighty hunter Orion (seen in the winter sky) by stinging him on the foot. To punish the scorpion for this dastardly deed, it was banished to the sky forever and placed directly opposite Orion so that it could never do him harm again.

Both Sagittarius and Scorpius are found along the ecliptic, the path in the sky followed by the sun, moon, and planets. In fact, this is probably how Antares got its name. "Ant-ares" literally translates from Arabic as "rival of Mars," a rivalry probably caused by the star's amazing resemblance to the red planet when seen with the naked eye.

Though there are twelve constellations assigned to the ecliptic (or zodiac), the sun will actually travel through thirteen each year. That forgotten constellation of the Zodiac is **Ophiuchus the Serpent-Bearer.** Ophiuchus fills most of the large void between Antares and Vega. Our ancestors drew a picture of the ancient world's greatest physician among these faint stars. Legend tells of how Ophiuchus (also known as Aesculapius, the son of Apollo) became such an amazing healer that Pluto, god of the underworld, complained about his lack of new clientele! Finally, to appease Pluto, Jupiter killed Ophiuchus and placed him among the stars as a lasting memorial.

Ophiuchus is shown holding a serpent named **Serpens.** Serpents and snakes have long been associated with healing for their periodic shedding of their skin, looked upon by our ancestors as a rebirth. This association of snakes and healing is still evident today on the symbol of the American Medical Association. Serpens is divided into two parts. Its head, shown on the map as Serpens Caput, extends to the right (west) of Ophiuchus, while its tail, Serpens Cauda, lies to the left, or east.

The next constellation encountered east of Sagittarius is **Capricornus** the Sea-goat. You have never heard of a sea-goat? A sea-goat is a creature with the head and front legs of a goat and the tail of a fish. It's odd that nowhere in classical mythology is such a bizarre animal mentioned. It has been suggested by some that Capricornus actually represents the mythical prankster Pan. Pan had the head and legs of a goat but the body of a man. The faint stars of the sea-goat form a large triangle seen in summer's southeastern sky.

Just above (north of) the sea-goat and to the left (east) of Altair lies the tiny but distinctive constellation of **Delphinus the Dolphin.** Delphinus is formed by a diamond of four stars composing the dolphin's body and a fifth star below (south) marking its tail. Not all cultures saw a dolphin among these stars. The ancient Chinese, for instance, saw a camel, with the diamond representing the camel's hump. Be it a dolphin or a camel, none of the stars in Delphinus are very bright. Still, its kite-like pattern stands out quite well against the barren surroundings.

Though the sun sets late on summer nights, this season's sky holds many treasures awaiting your watchful eye. Summer nights are made for stargazing.

6 Locating the Constellations and Bright Stars of Autumn

Level: All

Objectives:

- To find the prominent constellations visible in the autumn
- To learn the associated myths and legends

Materials:

- A clear autumn night sky
- A red-filtered flashlight (see Activity 2)
- A seasonal star map (Figure 6-1)

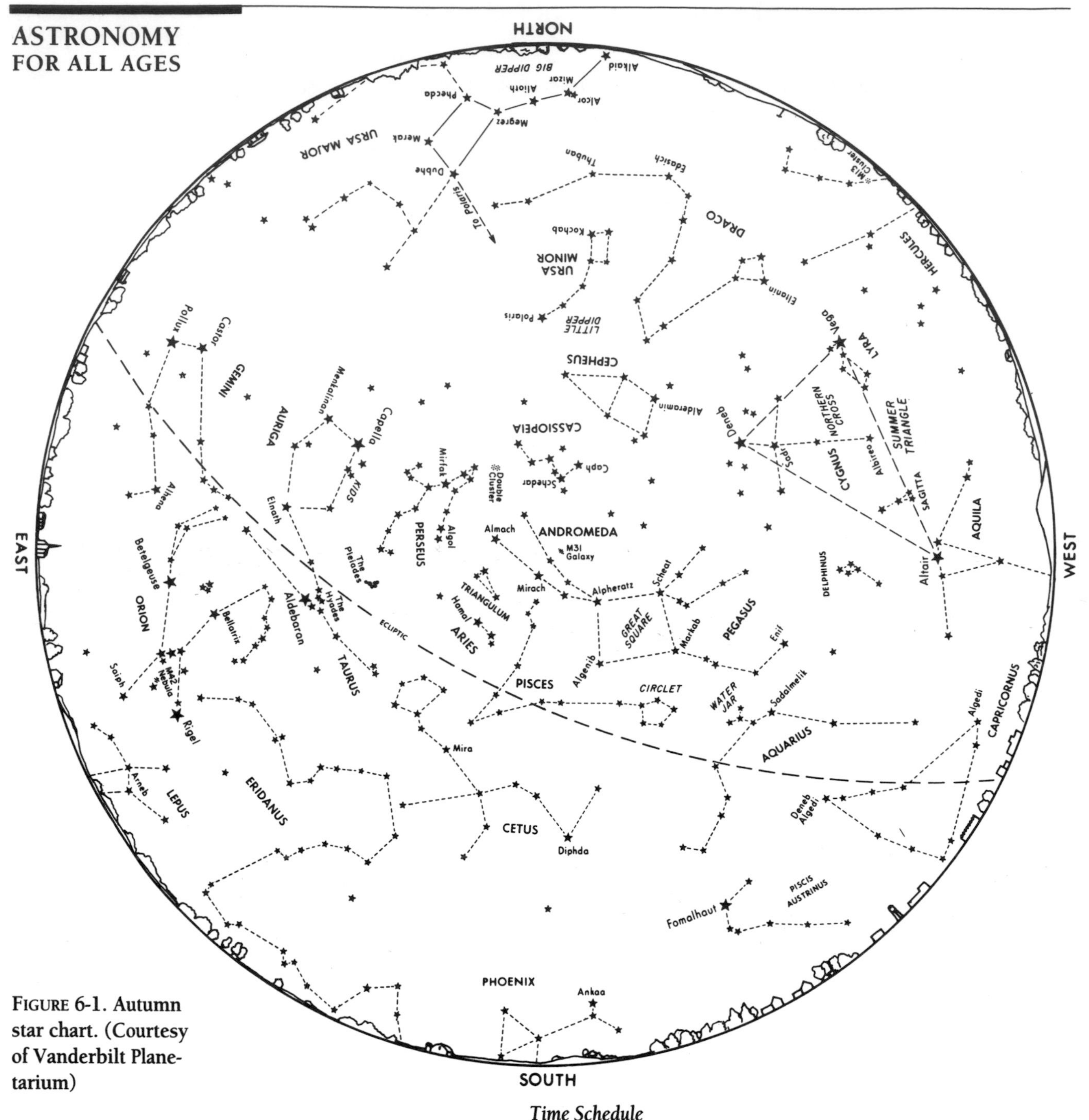

FIGURE 6-1. Autumn star chart. (Courtesy of Vanderbilt Planetarium)

Time Schedule

Early October: 12 Midnight	Early November: 10 P.M.	Early December: 8 P.M.
Late October: 11 P.M.	Late November: 9 P.M.	Late December: 7 P.M.

All times are standard. Add 1 hour for daylight saving time.

Both Earth and sky are in states of transition during the autumn months. As the countryside is painted in colors of red, orange, and yellow (no doubt the result of the wounding of Ursa Major, the Great Bear, by its pursuing Indians; see Activity 3), the night sky slowly turns away from the brilliant stars of summer to an area of fewer dazzling celestial jewels. Yet the autumn sky is a fascinating vista for stargazers, as you are about to see.

Just as the summer sky was dominated by a large triangular pattern, the autumn sky is accented by another geometric shape—a square. Flying high in the southern sky are the four stars of the **Great Square** of **Pegasus the Flying Horse.** Trying to imagine a horse (let alone

one that flies) among the stars of Pegasus is a difficult task for our imaginations. Perhaps you can see him flying upside down, with the square representing the body. The horse's neck and head curve from **Markab** to **Enif,** its front legs extend above the square, but the tail end of the horse is nowhere to be found.

Since autumn is traditionally associated with baseball and the World Series, it might be easier to see a baseball diamond among the stars of Pegasus. Take **Scheat** as home plate, **Alpheratz** as first base, **Algenib** as second, and **Markab** as third. A faint star in between second and third (not shown on the map) might be the shortstop, while another faint, unplotted star in the center of the square represents the pitcher's mound. Can you find these in the night sky? A pair of dim points of light just southwest of Scheat might even be one of the managers arguing with the umpire. And just look at all the fans scattered across the sky!

Technically, the star Alpheratz does not belong to Pegasus but to the neighboring constellation **Andromeda the Princess.** Andromeda is formed by two lines of stars framing her body. Here we also find the most distant object visible to the unaided eye, the Andromeda Galaxy (M31 Galaxy on the map). We will revisit the Andromeda Galaxy in Activity 43.

Andromeda is a member of the Royal Family of Autumn that also includes King Cepheus and Queen Cassiopeia (see Activity 3). According to Greek legend, Cassiopeia was very attractive and very vain. She never tired of bragging about her great beauty. One day she boasted of being fairer than the sea nymphs, who were well-known for their exquisite beauty. The nymphs overheard this boasting and complained to their father, Poseidon, king of the seas. Poseidon became so infuriated that he created a flood and Cetus the Sea Monster.

Cetus was sent to devour the citizens of Ethiopia, the land ruled by Cepheus and Cassiopeia. King Cepheus was told that his people could be saved only if he were to sacrifice his daughter Andromeda to the sea monster. He had no alternative but to lead his daughter to the water's edge and chain her to a rock, left to the mercy of Cetus. But she was to gain a rescuer in the person of Perseus.

Perseus, the son of Zeus, had been ordered to kill a gorgon called Medusa, a very ugly creature whose hair was made of snakes. She was so ugly that anyone who looked at her face would turn to stone. To keep Perseus from that fate, he was given a highly polished shield and told not to look directly at Medusa but only at her reflection. He was also given a helmet that made him invisible and a pair of winged sandals that allowed him to fly. With these, Perseus was able to sneak up on Medusa, decapitate the horrible creature, and put her head in a leather bag. Some of Medusa's blood fell into the water to create Pegasus the Flying Horse. Perseus climbed on the back of Pegasus and flew away.

Meanwhile, back at the seashore, things were looking pretty grim for Andromeda. Perseus heard her cries for help and swooped down to rescue her. Telling Andromeda to close her eyes, he pulled Medusa's head from the bag and dangled it in front of Cetus, who instantly changed to stone. Perseus and Andromeda fell in love. They climbed on the back of Pegasus and flew off into the sunset.

In our autumn sky, **Perseus** can be found towering protectively over Andromeda. His stars form what looks to some as a badly bent boat anchor. From these, our ancestors imagined a battle-clad warrior holding a sword in one hand and the awful head of Medusa in the other. An interesting star in this head is **Algol,** which means "demon." This star, marking the winking eye of Medusa, varies in brightness every three days. These variations can be easily observed with the naked eye. Another interesting sight visible to a keen eye on a clear night is the Double Cluster of stars near the pointed top of Perseus's helmet. The Double Cluster will be revisited in Activity 41.

Cetus the Sea Monster (sometimes referred to as a whale) also belongs to the autumn sky. Found looming below the Great Square, the many faint stars of Cetus make the sea monster a difficult fish to catch from urban and suburban skies. Its brightest star, **Diphda,** rides so low in the sky that it may be blocked from view by trees or other obstructions. Even more interesting is the star **Mira,** which means "wonderful." At times Mira will shine about as brightly as Polaris, while at others it will be nowhere to be found. It takes Mira more than a year to complete one full cycle from bright to dim and back to bright again. Mira and Algol are two examples of what astronomers call variable stars—stars that change in brightness.

Skimming autumn's southern horizon is **Fomalhaut,** brightest star in the otherwise-obscure constellation of **Piscis Austrinus,** the Southern Fish. Above Fomalhaut are many dim stars that our ancestors assigned to **Aquarius the Water Bearer**. Here we find the figure of a man carrying a large jar on his back. In most depictions he is shown flooding the sky, with the water flowing through Capricornus, Piscis Austrinus, Cetus, and Pisces.

Like Aquarius, **Pisces the Fishes** is a zodiacal constellation. Except for that distinction, little attention would be paid to Pisces because its stars are so faint. The constellation is usually drawn as two fishes (one represented by the circlet of stars shown on the map) joined at the tails by a long rope. Sounds fishy!

East of Pisces along the ecliptic lies the small constellation of **Aries the Ram.** It contains only one bright star, **Hamal,** which combines with two fainter suns to form the head of the animal. Aries was the legendary Ram of the Golden Fleece.

When the ram died, its valuable golden fleece was placed in a sacred grove guarded by a ferocious dragon. Wanting to return the valuable golden fleece to its home country of Thessaly, Jason, heir to the country's throne, set out to find it. To help him in his quest, Jason gathered some of the mightiest warriors the world ever knew, among them Hercules, Orpheus, Castor, and Pollux. The ship they sailed on was named Argo, and its crew became immortalized as the Argonauts. Together Jason and the Argonauts fell into some high adventures but never found the fleece.

Though we do not find Jason in the stars, many of the Argonauts as well as portions of their ship are scattered across all four seasonal skies. In the autumn we might still be able to make out Hercules as it sets in the west. Orpheus's lyre, Lyra, is also getting progressively lower in the west. Both are constellations of summer. Rising in the east are the twin brothers Castor and Pollux, the Gemini twins. Their tale will be found in Activity 7.

Autumn is a season of change, both here on Earth as well as up in the sky. Enjoy the relatively warm nights now, for the cold of winter is just around the corner.

7 Locating the Constellations and Bright Stars of Winter

Level: All

Objectives:

- To find the prominent constellations visible in the winter
- To learn the associated myths and legends

Materials:

- A clear winter night sky
- A red-filtered flashlight (see Activity 2)
- A seasonal star map (Figure 7-1)

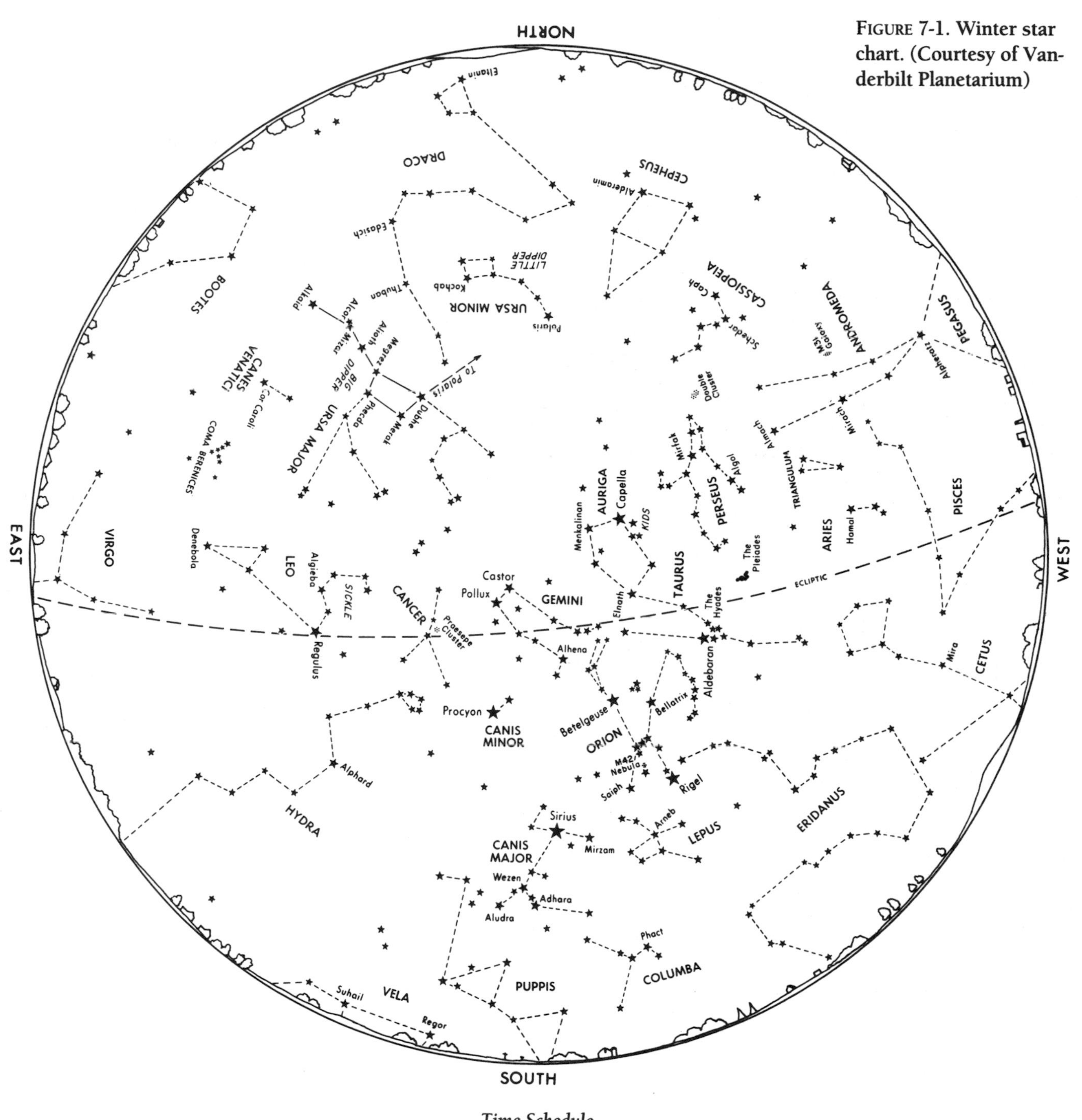

FIGURE 7-1. Winter star chart. (Courtesy of Vanderbilt Planetarium)

Time Schedule

Early January: 12 Midnight	Early February: 10 P.M.	Early March: 8 P.M.
Late January: 11 P.M.	Late February: 9 P.M.	Late March: 7 P.M.

All times are standard.

Ahh, the winter! For many stargazers, this is their favorite season. With the cleansing breath of a frigid arctic high-pressure weather front, the smog and haze that mask the heavens give way to crystalline nights. On those special nights, the stars seem to hang so close you can almost touch them! The sky of winter is glittering with a king's ransom of stellar jewels. Though it is sometimes difficult to face the cold temperatures of a winter's night, the sky, more spectacular than in any other season, makes it all worthwhile.

FIGURE 7-2. The prominent winter constellation Orion the Hunter.

Orion is the best constellation for demonstrating that stars come in different colors and that these colors depend on the stars' temperatures. Most people are aware that flames of different temperatures glow in different colors. A yellow candle flame, for instance, is cooler than the blue flame of a gas stove. The same is true of stars. The sun is a yellow star and has a surface temperature of about 11,000°F. Compare that to blue-white Rigel, in Orion. Rigel is one of the hottest stars known, with a surface temperature about 23,000°F! At the other end of the scale is red Betelgeuse, also in Orion. Its surface temperature is only about 5,000°F.

Standing center stage is **Orion the Hunter**, the most brilliant constellation of all (Figure 7-2). Look for three equally bright stars in a row, forming his belt. To the north lies the bright reddish star **Betelgeuse**, the right shoulder of the hunter (seen on our left). His left shoulder (on our right) is represented by the star **Bellatrix.** Above the shoulders is a faint group of three stars depicting Orion's tiny head. Below the belt are the stars **Rigel**, representing Orion's left knee, and **Saiph**, marking his right knee. He is shown holding a shield of faint stars in his left hand and a club raised high over his head in his right.

Hanging from Orion's belt are the dim stars of his sword. The middle star in the sword is not actually a star but a cloud of glowing hydrogen gas called a **nebula.** Astronomers have catalogued this nebula as M42, but most backyard stargazers know it better as the Great Orion Nebula (see Activity 42.)

According to legend, Orion was the mightiest hunter of all time. One day he boasted of being able to defeat any animal on earth. His constant bragging was overheard by Mother Earth who, fearing that he would destroy every creature, sent a scorpion to sting Orion on the heel and kill him. Diana, goddess of the hunt, honored the slain Orion by placing him among the stars. (The scorpion, Scorpius, was also placed in the sky. It may be found under heavy guard in the summer sky, directly opposite Orion so that it can never harm him again.)

Even in the sky, we find Orion doing battle again, this time with **Taurus the Bull.** To locate Taurus, draw a line through Orion's belt and extend it to the upper right (northwest). You will come to the bright orange star **Aldebaran,** marking the bull's eye. The head of Taurus is formed by a V-shaped group of stars nicknamed the **Hyades.** Imaginary lines extend to two stars above its head to create Taurus's long horns.

The story goes that Orion is trying to club the bull over the head in order to save seven sisters who were kidnapped by Taurus. We can still see the sisters trapped in the sky, formed by the tiny cluster of stars known as the **Pleiades.** Most people can spot six or seven stars in this region, though on an extremely clear night, some can see up to fifteen. A telescope or binoculars will reveal that nearly 100 stars form the Pleiades cluster (see Activity 41).

Though you cannot suspect it just by looking with your eyes, Betelgeuse is one of the largest stars ever discovered. Classified as a red supergiant star, Betelgeuse measures about 640 million miles in diameter, more than 700 times larger than the sun. If Betelgeuse were at the center of our solar system, its outer edge would extend beyond the orbit of Mars. (Mars, the next planet after Earth, is 141 million miles from the sun. Earth, at 93 million miles out, would be inside the star!)

Returning to the belt of Orion, continue the line down toward the southern horizon and **Sirius,** brightest star of the nighttime sky. Sirius is also known as the Dog Star, as it belongs to the constellation **Canis Major,** the Large Dog. Can you see a dog among its stars? Take Sirius as a jewel on the dog's collar. The star **Mirzam** marks a front paw, **Adhara** a hind paw, and **Aludra** the tip of the hound's tail.

Canis Major is but one of Orion's two faithful companions. His small dog, **Canis Minor,** can be found to the northeast. Canis Minor is made up of only two easily visible stars, the brighter called **Procyon.** Most people cannot actually see a dog here—other than perhaps a hot dog.

Watching Orion and Taurus battle are the twin brothers **Gemini.** Their heads are marked by the bright stars **Castor** and **Pollux,** with their stick-figurelike bodies extending westward. Castor was a famous horse trainer and soldier, while Pollux is remembered as Sparta's leading boxer. Ancient legend tells of the heroism of Castor and Pollux. For instance, they were members of the Argonauts, the legendary crew of the ship Argo. Led by Jason, the Argonauts searched for the golden fleece, encountering adventure at every turn.

The twins appear to be standing on the hazy band of our Milky Way galaxy. Not as bright as the summer portion, the winter Milky Way nonetheless contains many fascinating sights when searched with binoculars or a telescope.

Following the Milky Way northward, we find the constellation **Auriga the Charioteer.** The bright star **Capella** marks his left shoulder; dimmer **Menkalinan** represents his head. While Auriga is commonly referred to as a charioteer, he is frequently drawn as a shepherd holding three young goats. The major stars in Auriga form sort of a celestial home plate or pentagon.

By connecting the stars Capella, Castor, Pollux, Procyon, Sirius, Rigel, and Aldebaran by a curved line, a giant circle is formed. This is the **Winter Circle of Stars,** dominating the evening sky from December through March.

Some lesser-known constellations also populate the winter sky. For instance, directly below Orion is the faint but easy-to-spot constellation **Lepus the Hare.** The body of the hare is formed from a trapezoid of four stars, with the star **Arneb** marking its nose. Two stars form a northward-curving arc with Arneb to create the animal's long ears, while two other stars to the west of the trapezoid might be the ends of a carrot dangling in front of Lepus.

Even more difficult to spot is **Eridanus the River.** Composed mostly of faint stars, Eridanus winds its way slowly toward and below the southern horizon. At the "mouth" of this celestial river lies **Achernar,** one of the brightest stars in the night sky. Unfortunately its position in the far southern sky prevents it from being seen from all but the extreme southern tier of the United States.

Though the cold temperatures outside are challenging, the warm and wonderful stars and constellations make it worth bundling up to visit with the winter night sky.

8 Which Stars Can I See Tonight?

Building Your Own Planisphere

Level: Intermediate, advanced

Objective:

- To make a device that shows what stars are visible at any time of the night and any night of the year

Materials:

- Full-size photocopies of Figures 8-1 and 8-2
- Scissors
- Clear plastic report cover (discard the plastic spine)
- A piece of thin cardboard about 8 inches square
- A piece of corrugated cardboard about 10 inches square
- Rubber cement
- Cellophane tape
- Paper fastener

BACKGROUND

How did you do using the star charts in the previous activities? All were drawn to be used during their respective seasons, but their usefulness is somewhat restricted by the time of night. Each chart showed the sky for a certain hour and time of month, as specified along their bottom edges. But what if you weren't looking at the sky at those times? What if instead you wanted to look at the sky at, say, 11:00 P.M. on August 12? Or maybe 2:00 A.M. on December 13? What stars would you see then?

To answer those questions, many stargazers use "adjustable" star charts called **planispheres.** A planisphere may be used to match the exact month, day, and hour when you are looking at the sky. How does such a magical device work? Very simply!

Take a look at Figure 8-1. Called a star wheel, it shows the major stars and constellations across all four seasons. The North Star is marked by the boxed X at the center of the star wheel. You will probably recognize many of the constellation names from the previous activities. Encircling the outer edge of the wheel are the twelve months, each broken into three sections: early, mid, and late.

The second illustration, Figure 8-2, shows the horizon and the four compass points—north, south, east, and west. This is your "window" on the sky. By marrying these two drawings together, you can make your own planisphere. (Note that even though the sky is perceived as circular, the sky window must appear oval for the planisphere to work properly.)

ACTIVITY

Begin assembly of your planisphere by making photographs of Figures 8-1 and 8-2. Place the photocopy of the window (Figure 8-2) in the plastic report cover. Using a Magic Marker

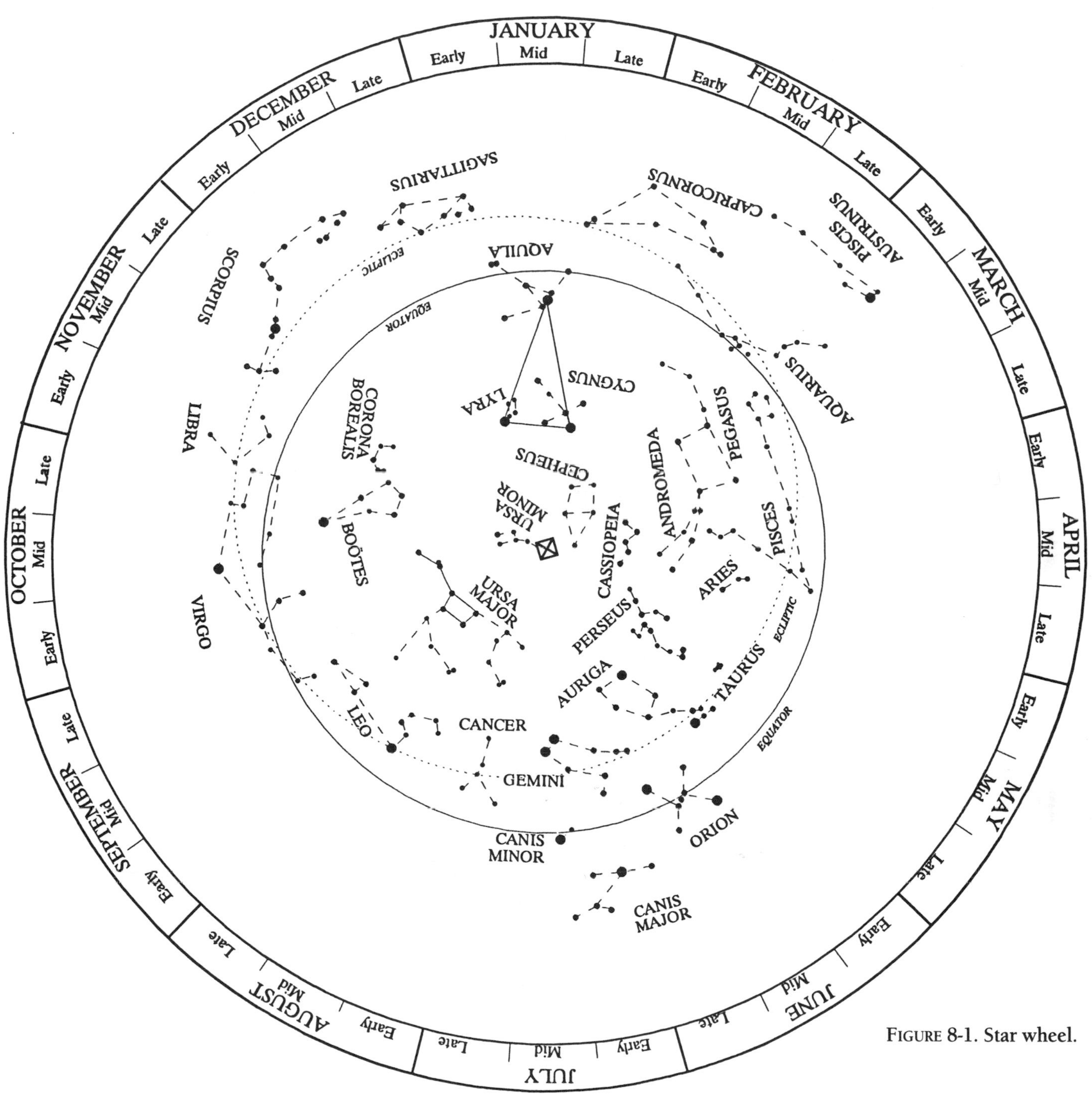

FIGURE 8-1. Star wheel.

or pencil, mark the exact location of the window's X on the front of the plastic folder. Remove the window photocopy and carefully punch a small hole through the X and through both sides of the report cover.

Glue the window sheet (Figure 8-2) onto an oversized piece of cardboard using rubber cement. When the cement has set, cut out the inner shaded oval window from the paper. Be sure not to slice into the "horizon." (For this step, you may prefer using a single-edge razor blade instead of scissors. If so, be very careful.) Discard the shaded area, but keep the background framing. Also cut out around the window framing.

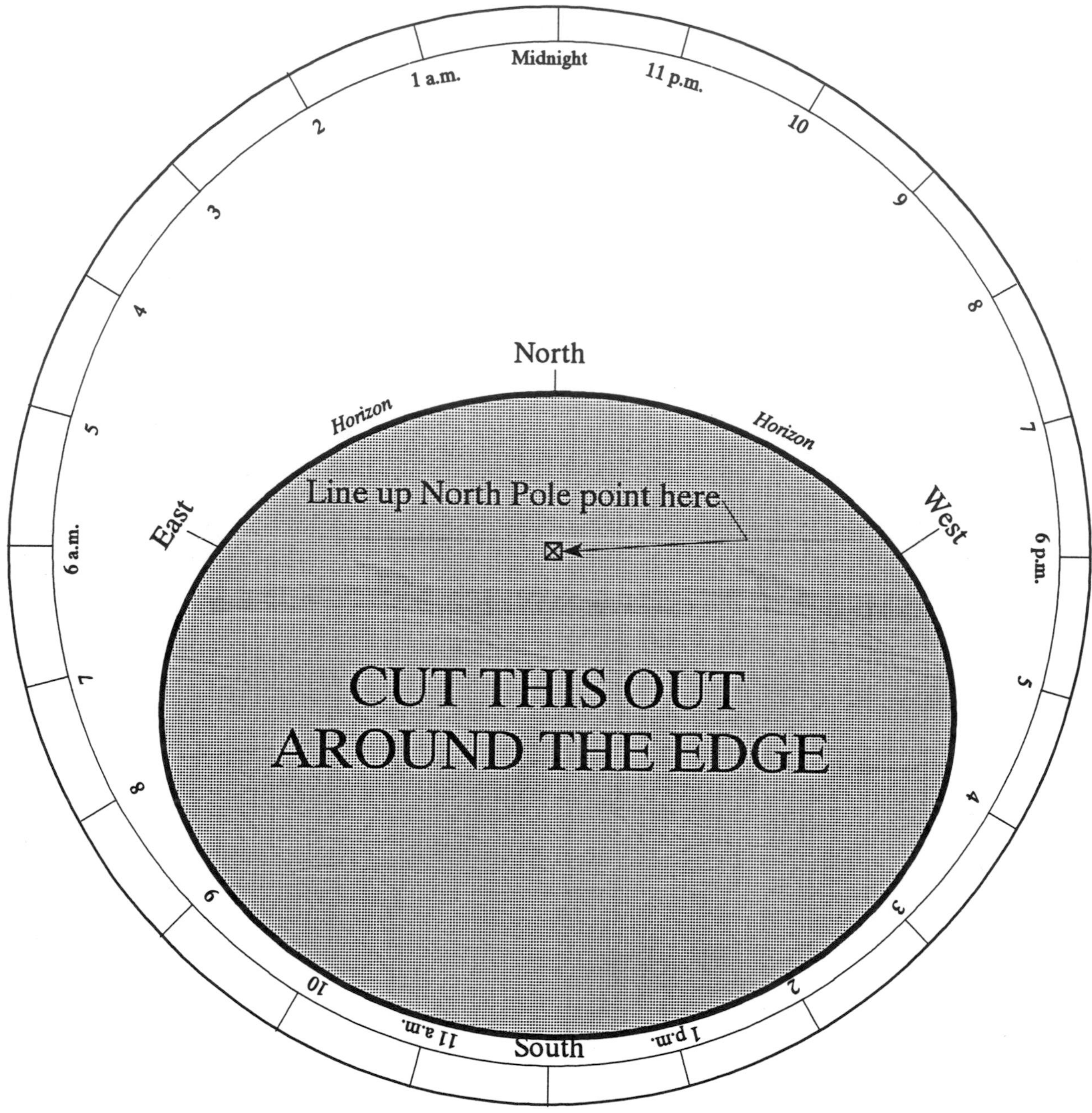

FIGURE 8-2. Sky "window."

Next, glue the star wheel (Figure 8-1) onto an oversized piece of corrugated cardboard. When dry, cut around the edge of the star wheel. Finally, using a pencil, punch a small hole in the exact center of the wheel, piercing the center X.

Place the window framing back in the clear plastic report cover and trace its outer diameter onto the cover. Cut out a circle from the plastic cover one inch larger in diameter than the window framing. Tape the window to the front (and only the front) of the folder. Tape only around the edges. Slip the star wheel behind the window, lining up the hole in the wheel with that in the cover. Stick a paper fastener through the holes in the front and back of the report cover and the star wheel. Bend back the ears of the fastener to hold it in place. You're done.

The only difference between using the planisphere and using one of the previous star maps is that it must be set for the right time and date. To do this, simply match up the time of month found along the edge of the star wheel with the hour found along the edge of the sky window. When properly set, the planisphere will show the stars visible at that time and date. In no time you'll be hunting stars and constellations at all hours of the night.

9 Star Brightness I

How Bright Is That Star?

Level. Intermediate and advanced

Objective: To learn how astronomers rate the brightness of stars

Materials: A clear nighttime sky

BACKGROUND

It should be apparent even to the casual observer that all stars in the night sky do not shine with the same intensity. Some appear quite bright, others are barely discernible, while most are somewhere in between.

When astronomers talk about how bright something appears in the sky, they refer to that object's **magnitude.** Over 2,000 years ago the Greek astronomer Hipparchus devised a system for categorizing stars according to their brightness. His method was quite simple. The brightest stars were labeled first magnitude, while the faintest stars visible to the eye were sixth magnitude. The remaining stars fell somewhere in between. In the Hipparchus system, the larger the magnitude number, the fainter the star. This idea always confuses people at first, but don't worry, you'll get used to it.

The magnitude system in place today is far more precise but still strongly reminiscent of Hipparchus's. Today the brightest objects have negative magnitude values, such as the sun at magnitude –26.7. With the invention of the telescope in the early seventeenth century, the magnitude scale also had to be extended the other way. Telescopes and binoculars can reveal stars much fainter than sixth magnitude.

Compare Figure 9-1 with Figure 7-2 in Activity 7. See how many more stars are shown in Figure 9-1. Figure 7-2 shows the winter constellation Orion as it appears to the naked eye from suburban skies; that is, the photo records stars to about fifth magnitude. But Figure 9-1 reveals stars as faint as ninth or tenth magnitude, about the same as can be seen in small telescopes. (This photo records more stars because the film was exposed for a longer time.) The largest professional instruments exceed twenty-third magnitude.

By contemporary standards, a one-magnitude jump (say from first to second, or second to third) corresponds to a change in brightness of 2.5 times. Therefore a first-magnitude star is 2.5 times brighter than a second-magnitude star, while a second-magnitude star is 2.5

FIGURE 9-1. Compare this photograph of Orion to Figure 7-2. Note how many more faint stars are visible in this picture. (A longer exposure here recorded the fainter stars.)

times brighter than a third-magnitude star. By this method, a first-magnitude star is 6.3 times brighter than a third-magnitude star (2.5 x 2.5 = 6.3), and so on. A five-magnitude jump, say from first to sixth magnitude, matches a change in brightness of 100 times.

Table 9-1 lists the brightest objects in our sky and their approximate magnitude values.

TABLE 9-1

The Brightest Objects in Our Sky

OBJECT	MAGNITUDE
Sun	–26.7
Full Moon	–12.7
Venus	–4.7
Jupiter	–2.9
Mars	–2.8
Sirius	–1.4
Arcturus	–0.3
Saturn	0.0
Vega	0.0
Capella	+0.1
Rigel	+0.1
Procyon	+0.4
Betelgeuse	+0.4

Note: The moon and planets can vary greatly in brightness depending on their position with respect to Earth. Mars, for instance, can be the fifth brightest object in our sky at times, while at others it appears as just another point of light in the sky.

ACTIVITY

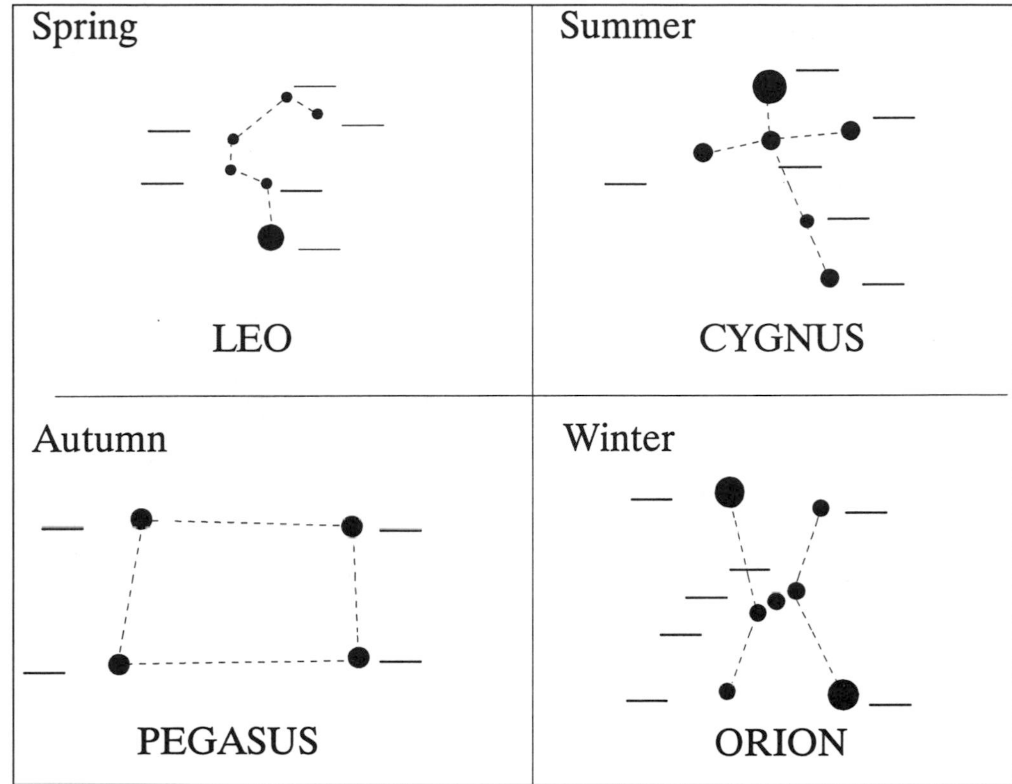

FIGURE 9-2. Rate the stars in each of these four seasonal constellations according to their brightness, noting the brightest star in each as "1," second brightest as "2," and so on. Then compare your results to Figure 9-3.

Figure 9-2 shows portions of four prominent seasonal constellations. You can find each of them on the appropriate seasonal star charts found in Activities 4 through 7. On the next clear night, go outdoors and find the star pattern in the sky that corresponds to the current season. Study the stars closely with just your eyes (no telescope or binoculars needed). Then, bringing a copy of the figure outdoors with you, rank the stars in descending order according to their brightness. Which star in the pattern do you see as the brightest? Mark the blank next to that star with the number 1. Which is second brightest? Mark that star number 2 on the chart, and so on. Continue until all blanks are filled in. You can always change the ranking, so don't rush.

Back indoors, turn to Figure 9-3 to see how well you did. You may find that your first attempt will be a bit off from this answer key. Try this activity in each subsequent season, however, and by the time a year has passed, your eyes will have developed as accurate measuring tools. The correct order and magnitude values are given in Figure 9-3—but no peeking ahead of time!

Although many people erroneously refer to a bright star as "large," size has nothing to do with its apparent brightness. Some of the brightest stars in the sky are much smaller than other, fainter stars. As we will demonstrate in the next activity, a star's brightness in our sky depends on its size, its inherent luminosity, and its distance from us.

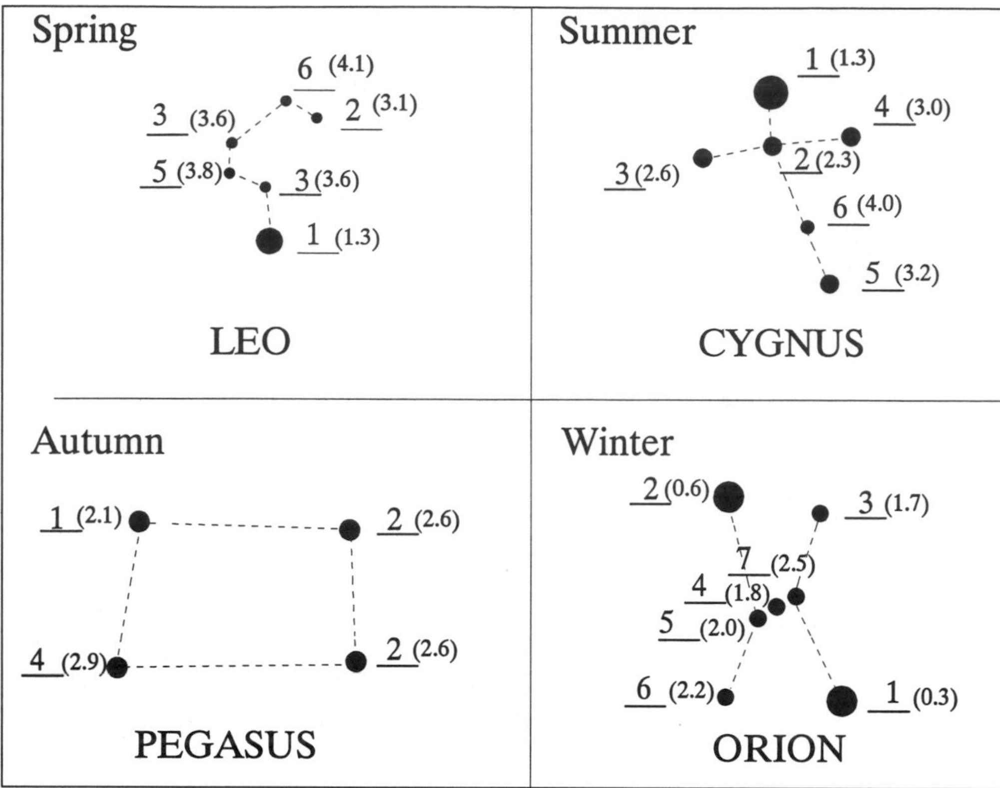

FIGURE 9-3. Answers to Figure 9-2. Exact magnitude values for each star are shown in parentheses.

10 Star Brightness II

Don't Judge a Book by Its Cover

Level: Intermediate and advanced

Objective: To show that a star's magnitude in our sky depends on both its intrinsic brightness and its distance from us

Materials: Two flashlights (one noticeably brighter than the other)

BACKGROUND

The last activity explained how astronomers rate the brightness of stars by a system of numbers called magnitudes. But *why* do some stars appear brighter than others in the night sky?

When asked this question, most people will answer, "Because they're bigger." While this is usually true, it is not the entire answer. (Surprisingly, some of the largest stars ever found are among the faintest. But these are the exceptions to the rule.)

The complete answer to this question comes in two parts. A star may appear bright in our sky either because it is intrinsically very bright or because it is close to Earth. In this activity, we are going to prove that either one or both of these answers could be right.

ACTIVITY

a

FIGURE 10-1a (top), FIGURE 10-1b (below). Which flashlight is closer? In *a* both Helen Harrington (left) and Erin Logarzo (right) are the same distance from the camera, while in figure *b* Helen, with the brighter flashlight, is farther away.

This experiment requires two flashlights, a long, dark room or yard, and at least one helper. Make sure that one of the flashlights is brighter than the other. For this exercise we will refer to the brighter flashlight as a bright star and the dimmer flashlight as a faint star. Turn both lights on and have your assistant hold them the same distance from you. What do you see? You should notice that the flashlight representing the brighter star is brighter than the one portraying the dim star, as in Figure 10-1a. So far, no surprises.

b

Now have your helper place the dim flashlight on a table or other support (or have someone else hold it) and walk farther away from you, all the time aiming the bright light your way. As the distance between you and the bright star increases, what do you notice about its brightness? It should appear to be getting fainter and more like the brightness of the fainter, closer star. As the distance between you the observer and the bright star increases, its apparent brightness will equal that of the faint star, as in Figure 10-1b. Eventually, if space permits, the faint star will appear brighter than the bright star.

A good example of this situation in the real sky is the star Betelgeuse, the red shoulder star in the winter constellation Orion the Hunter. As mentioned in Activity 7, Betelgeuse is one of the largest stars known. If it were placed in the center of our solar system, the edges of the star would extend out beyond the orbit of Mars. Betelgeuse is thousands of times larger than our own star, the sun, yet it appears much fainter. Why? Simply because Betelgeuse is much farther away. The sun is only 93 million miles away, while estimates place Betelgeuse much, much farther away, at 550 light-years from Earth.

A LIGHT-YEAR is a unit of measure used by astronomers when they refer to the distances to the stars. One light-year—that is, the distance a beam of light will travel in 365 days—equals approximately 6 trillion miles. Therefore, Betelgeuse lies a little over 3.2 quadrillion miles from us, and it takes 550 years for its light to reach Earth. By comparison, it takes sunlight only about eight minutes to travel from sun to Earth.

The magnitude of a celestial object usually refers to how something *appears* in our sky. In these cases, to be perfectly correct, the magnitude value should be called the object's **apparent magnitude.** The apparent magnitude of a star is just the brightness of that star as seen from Earth. But this tells us nothing about its intrinsic brightness, how bright it really is. That was made clear in the example of Betelgeuse versus the sun.

To determine the inherent brightness of a star, astronomers refer to its **absolute magnitude.** The absolute magnitude of a star is simply its apparent magnitude as seen from a predetermined yardstick distance away. Astronomers chose the distance 32.6 light-years (192 trillion miles). How bright will each star appear from that distance? A general rule is that all stars whose distances from Earth are less than 32.6 light-years will have an absolute-magnitude value that is less than their apparent magnitude. Conversely, all stars that lie greater than this distance away will have absolute-magnitude values greater than their apparent magnitudes.

TABLE 10-1

Apparent versus Absolute Magnitudes: Some Examples

STAR	MAGNITUDE[1]		DISTANCE (IN LIGHT-YEARS)[1]
	APPARENT	ABSOLUTE	
Altair	+0.8	+2.2	16
Arcturus	0.0	+0.2	34
Betelgeuse	+0.5	−7.2	1,500
Deneb	+1.3	−7.2	1,500
Rigel	+0.1	−8.1	1,400
Sirius	−1.5	+1.4	9
Vega	0.0	+0.6	25

[1] Values taken from *Observer's Handbook*, Royal Astronomical Society of Canada, 1993

Which star listed in the table above appears the brightest in our sky (that is, which has the greatest apparent magnitude)? Which star appears the faintest of those listed (that is, which has the lowest apparent magnitude)? Hint: Remember that the lower the number, the brighter the star, and that negative numbers are lower than positive numbers.

How do those answers compare to their absolute-magnitude values? Which star has the greatest intrinsic brightness? Which star has the lowest? Notice that the stars whose distances are greater than 32.6 light-years have absolute magnitudes that are greater than their apparent magnitudes. Just the opposite is true for the stars closer than 32.6 light-years.

To make this point clear, use one of your two flashlights. Have a helper turn the light on and hold it, say, 15 feet away. Now instruct your helper to move toward you about 5 feet, keeping the light trained on you. What happens to the light's intensity? It increases as the distance decreases, right? Reverse things and have your helper move away from you until he or she is about 30 feet away. How bright is the light now? It should appear fainter than it was originally.

As you can see, a star's appearance in our sky depends on both its distance from the observer as well as its true brightness. Only by using special instruments attached to their telescopes can astronomers analyze the light from stars to determine their distances, sizes, and true (or absolute) magnitude.

Finally, here are the answers to our short star-quiz about the magnitudes in Table 10-1: The star listed in the table that appears the brightest in our sky is Sirius; the star that appears the faintest is Deneb; the star that has the greatest intrinsic brightness (absolute magnitude) is Rigel; the star that has the lowest absolute magnitude is Altair. How did you do?

11 Telling Time by the Stars

Level: Intermediate

Objective: To learn how to use the stars to determine the time of night

Materials:

- Photocopy of the star clock (Figure 11-1)
- Cardboard
- Rubber cement
- Paper fastener
- Scissors

BACKGROUND

Long before the invention of mechanical clocks and watches, our ancestors told time by looking at the sky. They did this by knowing the exact position of the sun, moon, and stars during each season of the year. We can re-create how our ancestors used the stars at night to tell time by making and using something we call a **star clock.**

As was pointed out in Activity 3, the North Star (Polaris) appears to remain fixed in the northern sky as nearby stars appear to circle around it. This effect is caused by the fact that Earth's North Pole happens to be aimed almost directly at Polaris. As the Earth spins on its axis, completing one rotation every twenty-four hours, stars near Polaris appear to trace small circles around the pole. (Stars farther away trace circles as well. These circles, however, are so large that they are interrupted by the horizon, causing the stars to appear to rise and set.)

Imagine Polaris is at the center of a huge clock. The stars near Polaris are on the face of that clock, turning **counter**clockwise ever so slowly like an hour hand. You can't actually see them move minute to minute, but if you look at their positions at, say, 8 P.M., then go back out a few hours later, you will see that their positions have changed compared to the horizon. We can use this fact to learn how to tell time by the stars.

ACTIVITY

First, make your own star clock by photocopying Figures 11-1a and 11-1b. Glue the photocopies to stiff pieces of cardboard with rubber cement. After the cement has set, carefully cut out both circles **exactly** as shown. Gently punch a small hole about the diameter of a pencil through the centers of both circles as well as through the center of the rectangle. (Note that the center marks the location of Polaris, the North Star.) Attach the circles, small one on top, to the rectangle using a two-pronged paper fastener. You now have a star clock!

To use your star clock, go out on the next clear evening and find Polaris and the Big Dipper. (Both will be toward the north. If you can't spot them, the appropriate seasonal star chart in Activity 4, 5, 6, or 7 should help. But remember, Polaris is a dim star.) Once they are found, hold the star clock up so that the horizon line is even with the real horizon. Turn the outer dial with your thumb until the correct date lines up with the "Month" arrow. Holding the outer dial with your thumb so it doesn't move, turn the inner dial until the picture of the Big Dipper matches the way it looks in the sky. Read the time using the arrow marked "Time" on the inner dial. If the star clock is set correctly, it should be very close to the current time. (Keep in mind that when Daylight Savings Time is in effect, the star time will be an hour behind clock time.)

With practice you will find that the star clock is an accurate way of estimating actual clock time. In fact, once you get to know the sky, you may not even need the star clock in order to make a good guess of the time. After all, it's "in the stars."

Star ★ Clock

MONTH

NORTH

FIGURE 11-1. Star Clock. (Artwork courtesy of Richard Sanderson)

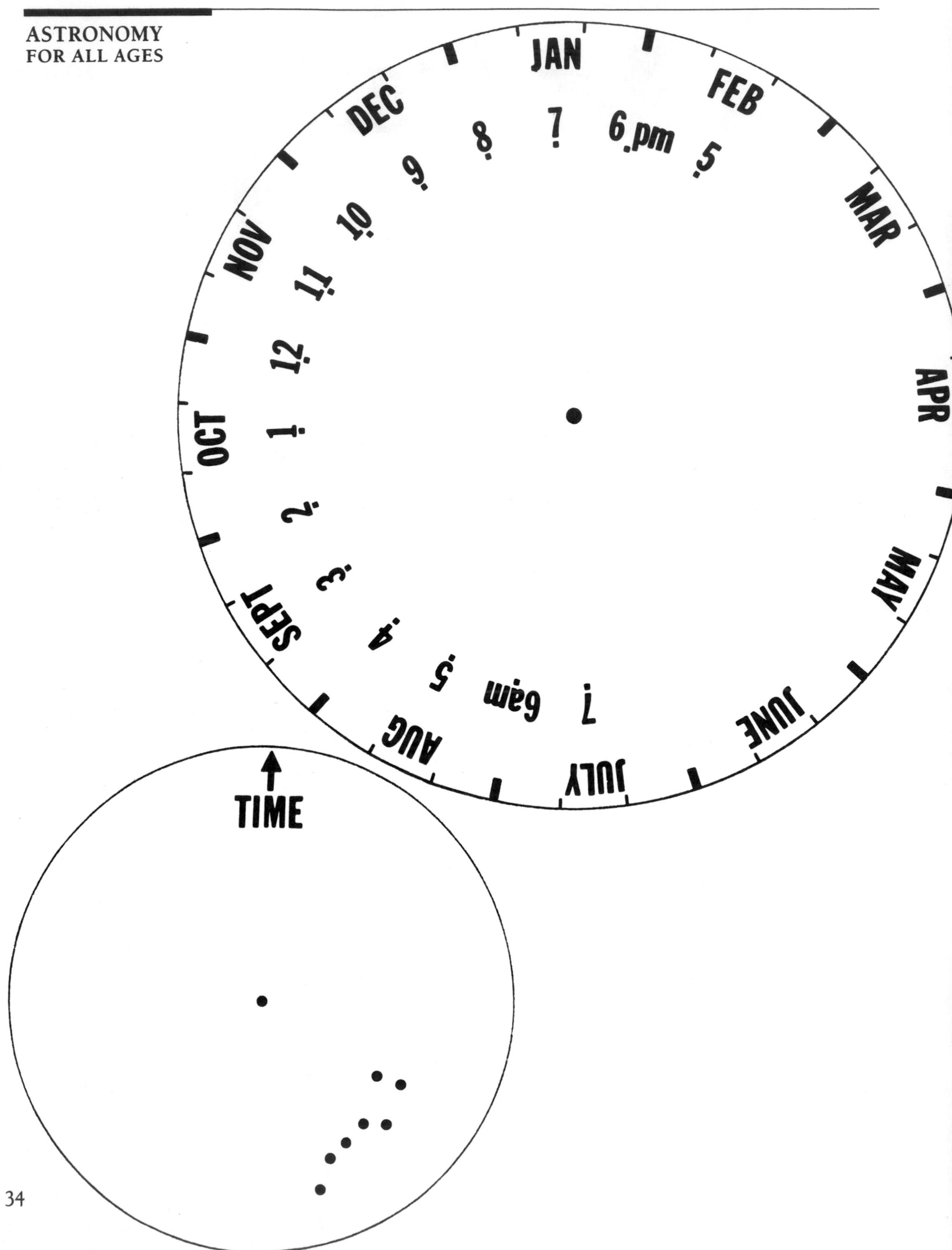
JAN
FEB
MAR
APR
MAY
JUNE
JULY
AUG
SEPT
OCT
NOV
DEC
7
6 pm
5
8
9
10
11
12
1
2
3
4
5
6am
7
TIME

12 Where Am I?

Using an Astrolabe to Find Your Latitude

Level: Intermediate and advanced

Objectives:

- To learn how to measure the altitudes of celestial objects
- To learn how to determine your latitude

Materials:

- An 8-inch drinking straw
- 8 to 10 inches of string
- A heavy nut, bolt, or key
- An 8-inch by 6-inch piece of cardboard
- Clear tape
- A straight pin
- A hole puncher
- A photocopy of Figure 12-3

BACKGROUND

An **astrolabe** is a simple, compact instrument that was used to calculate the positions of celestial bodies in the days before the invention of the sextant. It's still fun to learn how an astrolabe works and to see how much it can tell you. In the past, astrolabes were often elaborately decorated (Figure 12-1) and made of precious metals, like gold and silver.

FIGURE 12-1. A brass replica of a fifteenth-century astrolabe.

In this activity, you will sight on the North Star (Polaris) to determine your latitude—that is, your angular distance, in degrees, north of the equator. Christopher Columbus, during his voyages some 500 years ago, took many sightings of Polaris with an astrolabe in order to keep track of how far north of the equator he was as he traveled both east and west across the Atlantic Ocean.

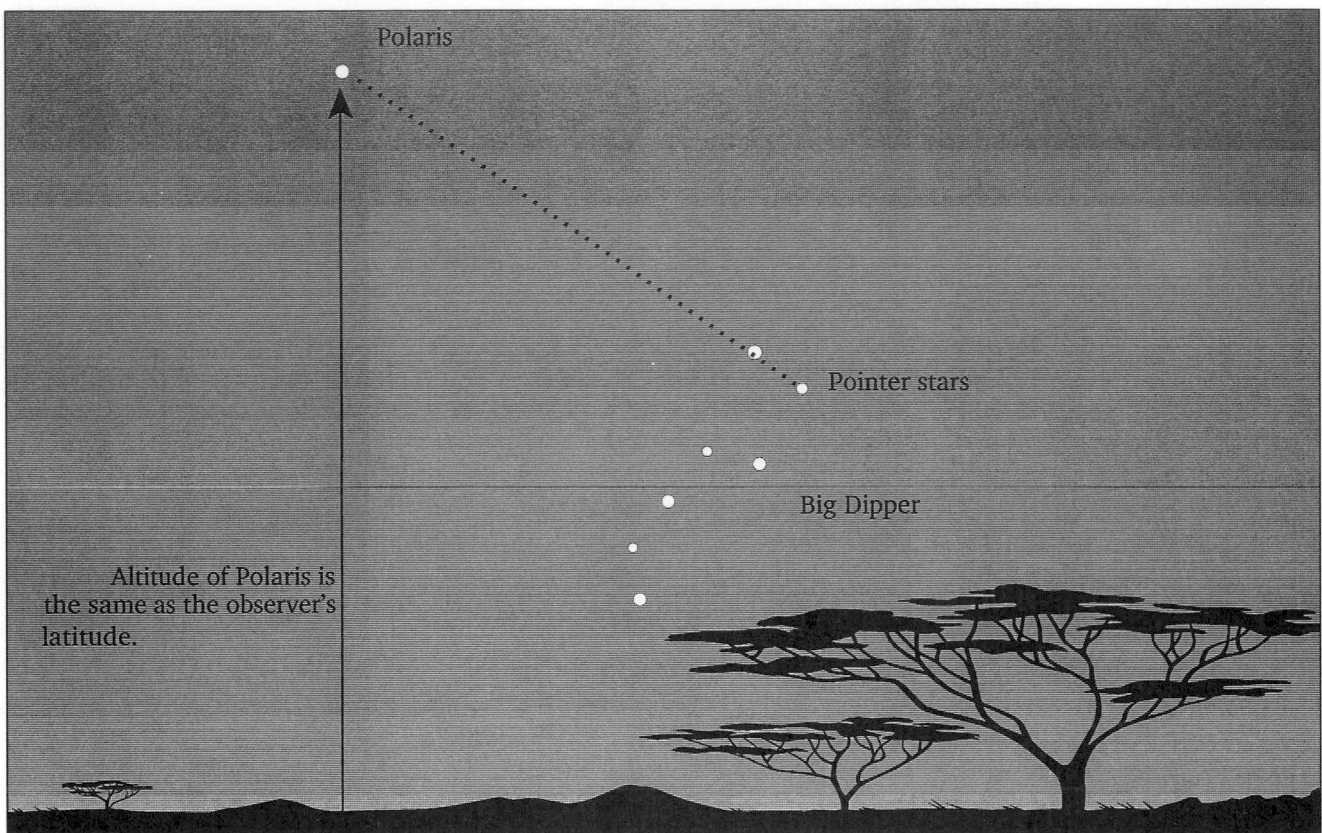

FIGURE 12-2. When measuring the altitude of Polaris, you will find that it equals your latitude on Earth.

The basic task of the astrolabe is to measure the angle of a star above the horizon. This angle is referred to as the star's **altitude.** As it happens, the altitude of the North Star equals the observer's latitude on Earth (Figure 12-2). This is because Earth's northern axis points nearly at the North Star, and the star thus remains fixed in the sky.

ACTIVITY

Photocopy and cut out the protractor in Figure 12-3. Punch out the hole shown in the protractor. Align the straight edge of the protractor with the 8-inch side of the cardboard (Figure 12-4) and tape the protractor to the cardboard. Push the straight pin through the cardboard at the center of the protractor hole and run the string through the hole; tie several knots in the end of the string so it won't pull back out. At the other end of the string, tie the nut or bolt so that it hangs freely when the cardboard is held vertically. Last, tape the drinking straw to the same edge of the cardboard that has the protractor's straight edge. You're ready to go!

Your astrolabe can now be used to measure the altitude of stars, planets, the moon—even the sun (with care, as described shortly). Go outdoors and locate the object you plan to measure. Hold the astrolabe up vertically, putting one end of the straw to your eye. Aim it until the object or star appears visible in the straw (but DON'T do this with the sun). Look-

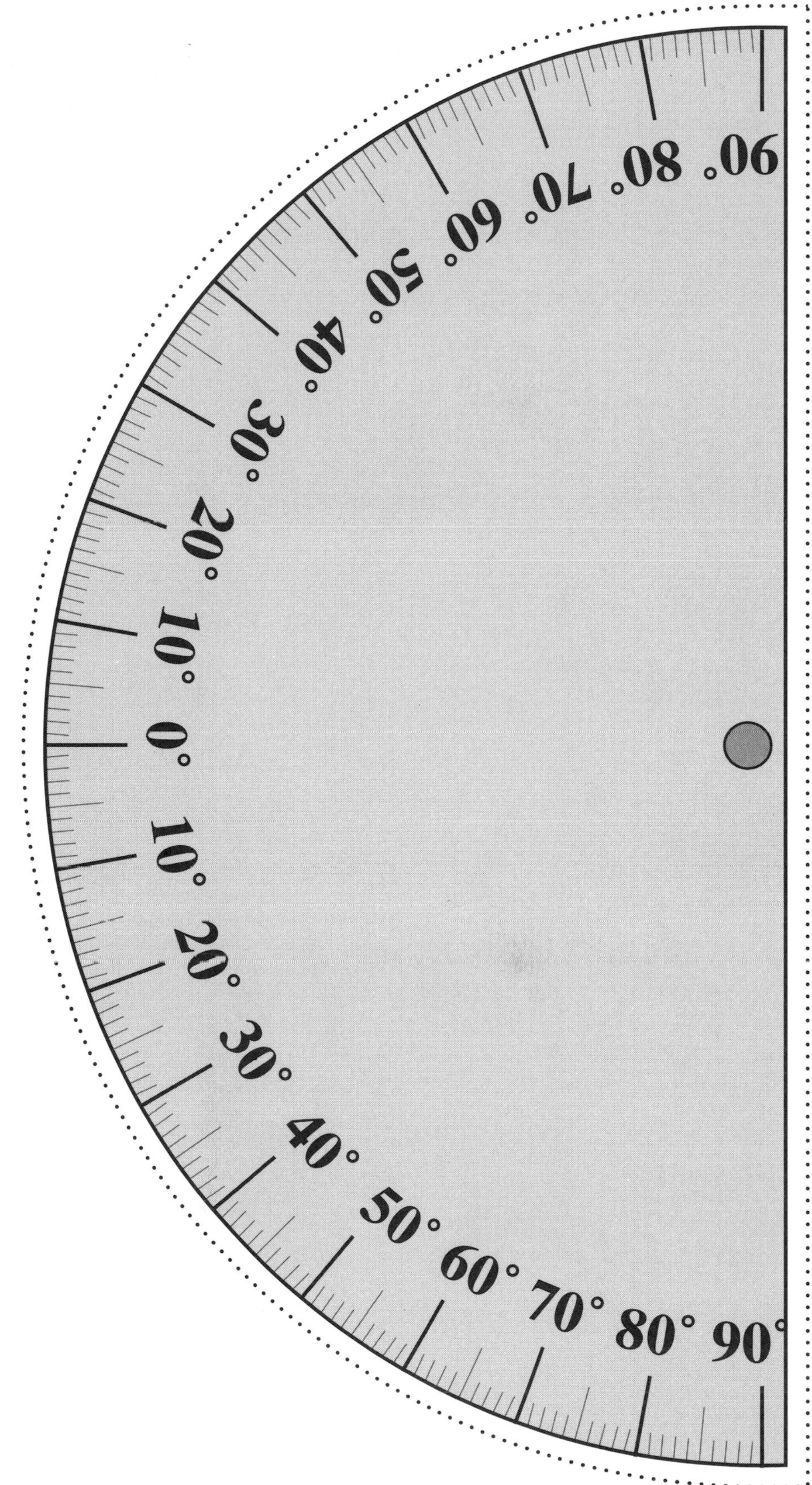

FIGURE 12-3. Protractor for the astrolabe. Photocopy, cut along dashed lines, and punch out circle at bottom.

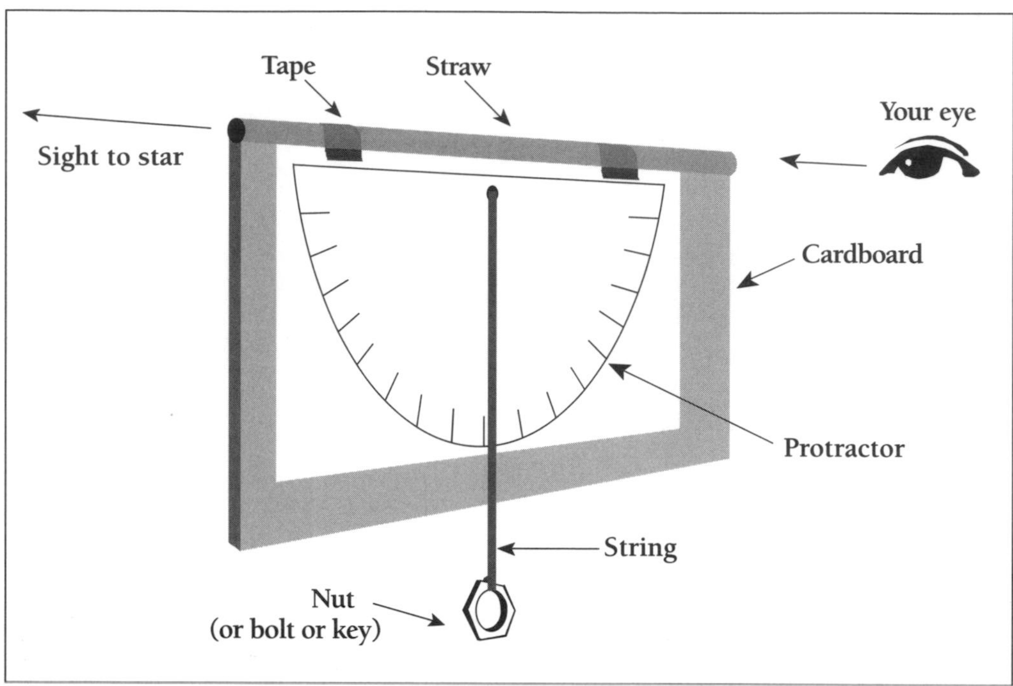

FIGURE 12-4. By sighting toward a star through the straw, the altitude of the star can be found by reading the angle behind the hanging string. You can thus find your latitude by sighting Polaris, the North Star, whose altitude remains fixed and is equal to your latitude.

ing through the straw is like using a tiny telescope. Hold the astrolabe very still with one hand as you use your other hand to hold the hanging weight against the cardboard. Then read the protractor angle where the string hangs; this is the angle (altitude) of the star above the horizon.

Try taking a sight of the North Star (Polaris), the one star whose altitude is very nearly fixed for any given location in the northern hemisphere. As we mentioned earlier, the altitude of the North Star equals the observer's latitude. Thus, by measuring its altitude, you can determine your latitude on Earth. Using your astrolabe, what is the altitude of Polaris? What is your latitude? Look up your latitude on a map of the United States in an **atlas** (a map book). Does the atlas agree with your astrolabe?

With the astrolabe, you can also follow the sun's changing daily and yearly altitude, but be careful! DO NOT LOOK AT THE SUN DIRECTLY THROUGH THE STRAW (see Activities 22 and 23 for tips on solar observing). Instead, face **away** from the sun. For the sun reading, you will **not** look through the straw. Point the straw toward the sun, but with your **back to the sun.** When the straw is aimed directly at the sun, sunlight will pass straight through the straw and create a small circle of sunlight on the ground. At this point, the straw's shadow will be its shortest.

Now, read the angle on the protractor where the hanging string touches it, and you've got the altitude of the sun for that moment of the day. Measuring the sun's noon altitude at the beginning of each season (see Table 12-1) helps show the relation between seasonal warmth, amount of insolation (whether direct or indirect rays of sunlight are falling on your location), and the altitude of the sun. At the beginning of which season will the sun's altitude be lowest? highest? After you have measured the altitude of the noon sun on the first day of fall, winter, spring, and summer, look carefully at the **difference** between each successive number.

At latitude 40° north, the noon sun's altitude on the first day of winter will be about 26.5°. On the first day of spring it will be 50.0°, for summer, 73.5°, and for fall, 50.0°. The difference between the sun's altitude at the beginning of winter and beginning of spring is

23.5°, and so on for spring and summer, summer and fall. This important astronomical number is the tilt of Earth's axis compared to the path of its orbit.

Once you're comfortable using the astrolabe, sighting these objects when you travel will give you a better appreciation of what our ancestors had to do.

TABLE 12-1

Using your astrolabe, measure the altitude of the following celestial objects. Also try choosing your favorite star and measuring its changing altitude through the course of one night.

OBJECT	TIME OF YEAR	ALTITUDE
North Star (Polaris)	Any	________
Midnight full moon	Midwinter	________
Midnight full moon	Midsummer	________
Noon sun	Vernal Equinox (March 21)	________
Noon sun	Summer Solstice (June 21)	________
Noon sun	Autumnal Equinox (September 21)	________
Noon sun	Winter Solstice (December 21)	________

13 "It Was This Big!"

Measuring Sizes in the Sky

Level: Intermediate and advanced

Objective: To learn how to measure celestial objects using angles

Materials:
- A clear view of the night sky
- Your hand

BACKGROUND

Imagine that on your way home one evening you see a bright meteor streak across the sky. In its short-lived passage, you notice that it left behind a long colorful trail. At home, in an attempt to describe this object and how far across the sky it appeared to travel, you find yourself at a loss for words. Instead, you stretch your arms out and say, "It was this long!"—just like the "fish that got away." Similarly, imagine a bright comet with a long tail. How much of the sky does the comet's tail cover? In astronomy, we ask "What is its **angular** size?" The angular size serves to tell how much of the sky is covered by an object. In this activity, you will learn how to approximate angular measurements in the sky in order to describe how large an object appears as viewed from Earth or how far apart certain objects are from each other.

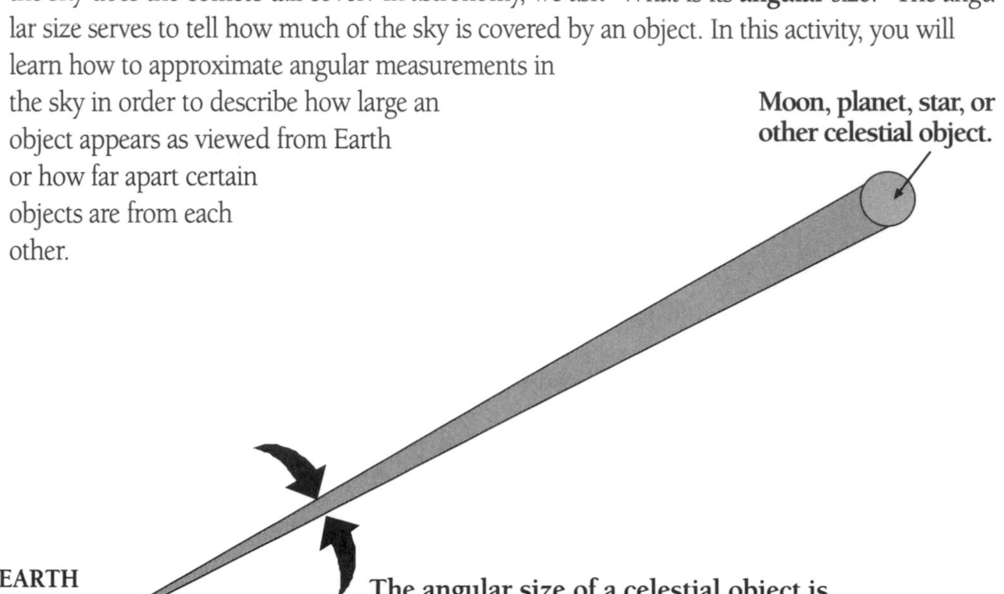

FIGURE 13-1. As seen from Earth, any celestial object will appear to cover a certain "piece" of the sky. The object thus has a certain angular size in the sky that depends on both its real size and its distance from Earth.

In astronomy, angular size refers to how large an object **appears** to be when viewed from a certain distance. You know that the farther an object is from you, the smaller it appears. This includes a person walking away from you, a distant building, cars on the ground as seen from an airplane—or celestial objects as viewed from Earth. The angular size of a celestial object simply means the number of degrees of angle that it occupies in the sky, as seen by you from Earth (Figure 13-1). Likewise, the angular **separation** of two celestial objects means the number of degrees of angle that separate the objects, as seen by you from Earth.

The entire sphere of stars around Earth (like a ball) contains 360 degrees (written 360°). Because you, an observer on Earth, see only half this amount of the sky when looking up, your entire sky covers only 180° (Figure 13-2). As you measure angular size or angular separation, you are actually figuring out how many degrees (out of the 180 visible to you) are covered by the objects you're viewing. For example, the stars Dubhe and Merak in the Big Dipper have an angular separation of 5°. You'll be able to make this measurement yourself later in this activity.

Astronomers use specialized instruments to measure angular size and separation. But you'll learn to do it by using your hand, held out at arm's length. In doing so, your hand will "cover" a certain piece of the sky. Although each of us has different-sized hands, the **proportion** of the length of our arms to the width of our hands is nearly the same for all people (try it on your friends). As a result, your hand may be used as a standard angular measuring device.

Try a test. To familiarize yourself with angular measure in the night sky, go outside and make fists with both hands. Stack one fist on top of the other one. Continue stacking fists, one on top of the other, until a fist points directly overhead (it should be about the ninth or tenth fist). Since it is 90° from straight in front of you (the horizon) to directly overhead (the zenith), each hand would thus represent an angular width of about 10°. Based on this, the open hand from thumb to pinky covers about 18° to 20° and an individual finger is about 1° across. Approximate angular sizes of the hand are shown in Figure 13-3.

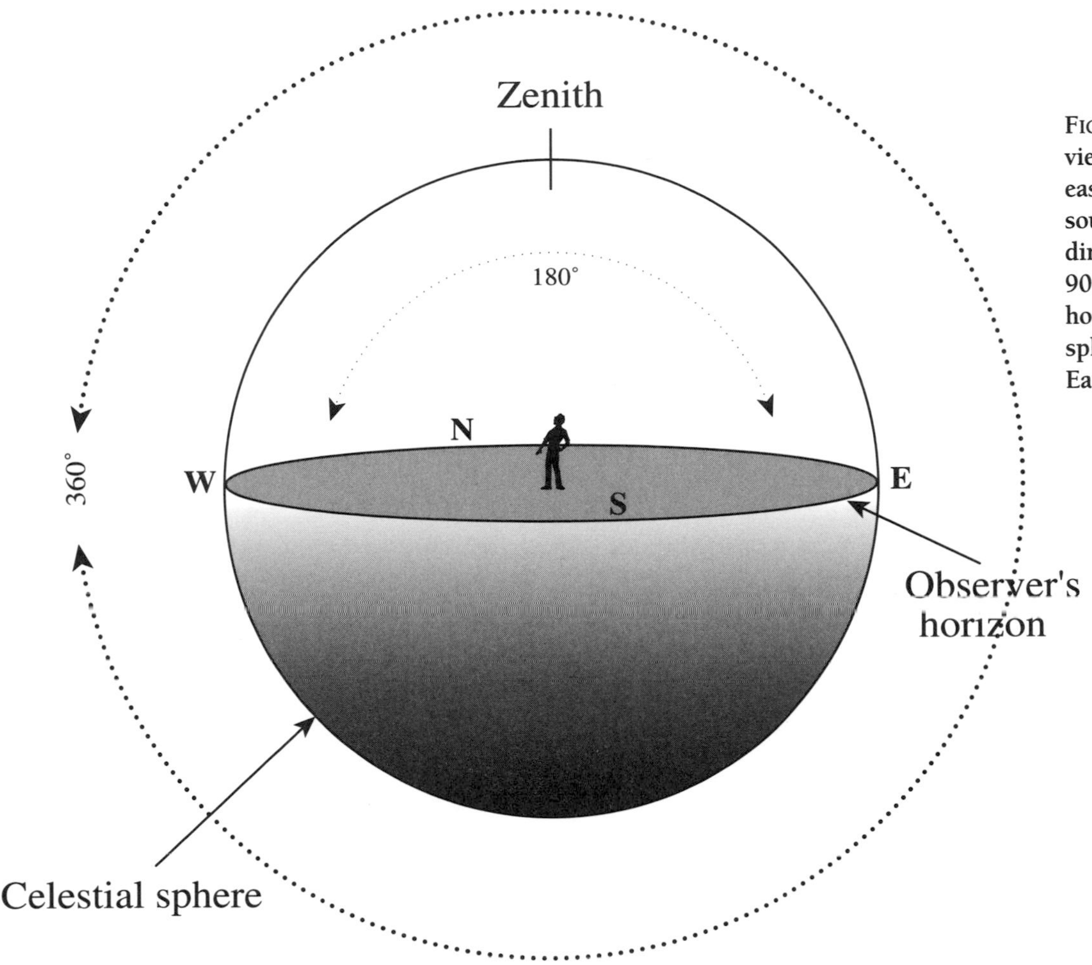

FIGURE 13-2. The observer's view of the sky, 180° from east to west and north to south. The zenith, the point directly overhead, marks 90° up and away from the horizon. The entire celestial sphere is 360° around Earth.

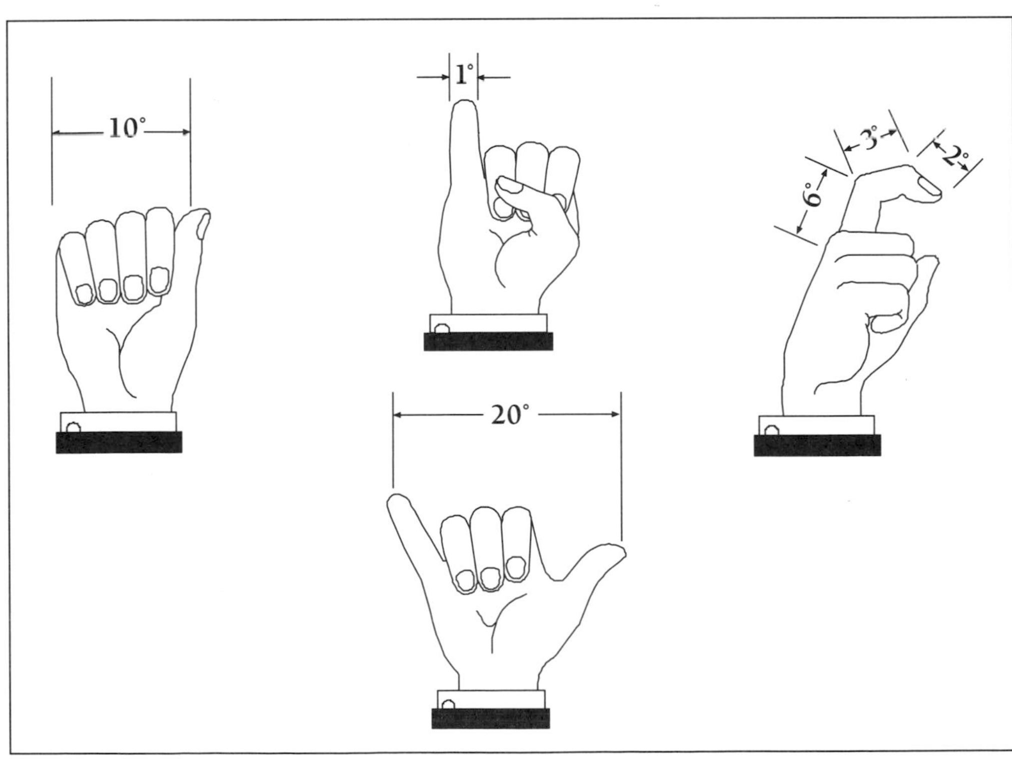

FIGURE 13-3. Approximate angular sizes of the hand when extended at arm's length.

ACTIVITY

Go outdoors on a clear night and practice measuring the angular size or separation of each of the objects in Table 13-1, using your hand. (If you like, you may also calibrate a 3-inch by 5-inch index card and hold it out at arm's length to take measurements. You will probably find that the long dimension of the card covers about 10°.) The objects in the table are listed by season so that they will be conveniently placed overhead within a few hours after sunset. Once you are more familiar with using your hand as a measuring tool, determining angular sizes in the sky will become second nature and thus the size of meteor paths and of other celestial scenery will be easy to measure.

TABLE 13-1

Using your hand as a measuring device, determine the angular extent of each of the following objects. Consult the seasonal star charts in Activities 4 through 7 to locate stars.

OBJECT(S)	SEASON FOR BEST OBSERVATION	ANGULAR SIZE OR SEPARATION	
		YOUR MEASUREMENT	ACCEPTED VALUE
1) The angular width of a basketball held 20 feet from you	Basketball season	________	(3°)
2) The full moon	Any (see Appendix 3 for phase dates)	________	(½°)
3) The separation of the Big Dipper's pointer stars (Dubhe and Merak)	Spring (May–June)	________	(5°)
4) The separation of the North Star (Polaris) from the Big Dipper star Dubhe	Spring	________	(30°)
5) The separation of Arcturus from Spica	Spring (May)	________	(30°)
6) The angular "length" of each side of the Summer Triangle:	Summer (July–August)		
a) Deneb - Vega		________	(20°)
b) Vega - Altair		________	(30°)
c) Altair - Deneb		________	(35°)
7) The angular width of the Perseus Double Cluster Fall	(December–January)	________	(1°)
8) The angular width of the Pleiades star cluster	Winter (January–March)	________	*
9) The angular width of Orion's Belt	Winter (January–March)	________	(5°)

* Believe it or not, the Pleiades has an accepted angular width of 2°; about four full moons across!

14 Spotting Artificial Satellites

Level: All

Objective: To locate artificial satellites passing overhead

Materials:

- A clear, dark, and unobstructed view of the night sky
- Wide-field binoculars (optional)

BACKGROUND

On October 4, 1957, Soviet scientists launched Sputnik I, the first artificial Earth satellite. Traveling some 18,000 miles per hour, it orbited our planet every 95 minutes before plummeting back to Earth on January 4, 1958. Sputnik (the Russian word for "traveler") marked the beginning of the Space Age. In the United States, the newly formed National Aeronautics and Space Administration (NASA) began to develop projects ranging from basketball-size satellites to orbiting space stations.

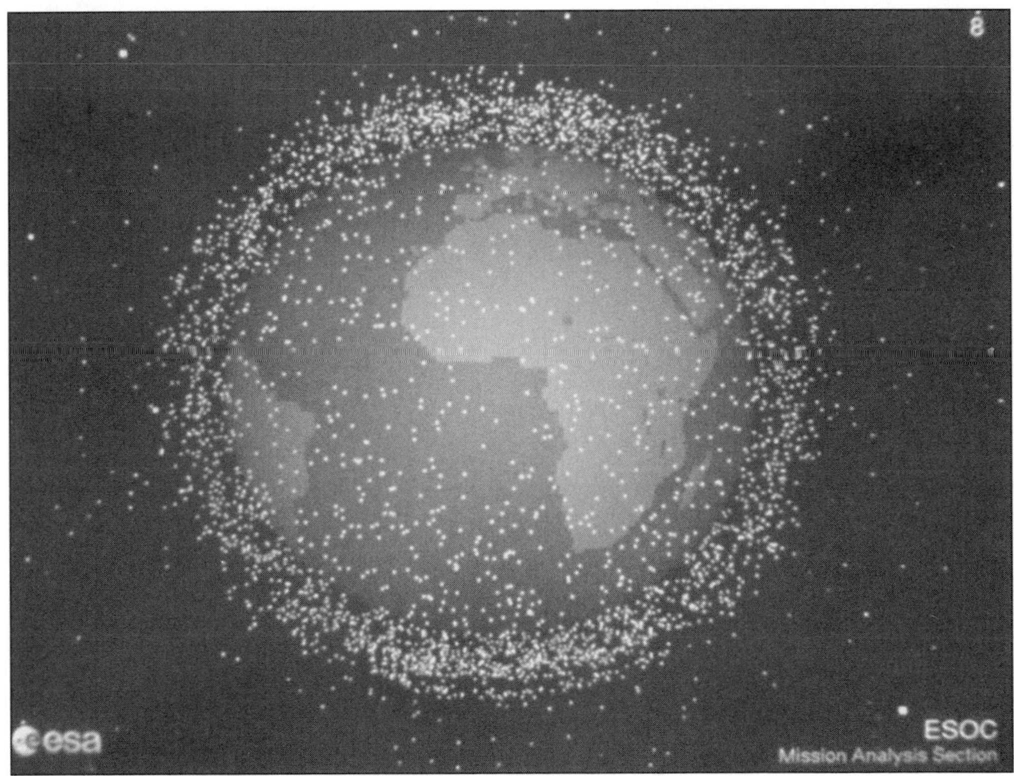

FIGURE 14-1. A computer plot showing active and defunct satellites as well as space debris up to an altitude of about 12,000 miles. (Photo courtesy of European Space Agency)

Nearly half of all functioning satellites lie at altitudes between about 21,000 and 23,000 miles. (For comparison, the moon orbits Earth at about 240,000 miles, while the space shuttle orbits at about 200 miles.) These high altitudes are often desirable because they put a satellite beyond most of the atmospheric friction that will slow it down and cause its orbit to decay more rapidly, thus decreasing the amount of time it can stay in orbit. The altitude of 22,300 miles is used for placement of **geosynchronous satellites.** A geosynchronous satellite is one that remains in place over a particular spot on Earth because the time it requires to orbit the planet precisely matches the rotation rate of the Earth (twenty-four hours). From its "stationary" perch in the heavens, such a satellite can continually observe the same place on Earth or relay communications to and from that place.

Satellites have altered our lives significantly. Can you think of several ways that satellites have had an effect in your life?

Some 400 functioning artificial satellites from nearly a dozen countries now orbit Earth. In addition, more than 7,000 cataloged objects (including 1,600 defunct satellites) and some 50,000 pieces of space debris (such as nuts, bolts, and metal scraps) orbit Earth (Figure 14-1). Placed in a variety of orbital heights, or altitudes, the types of functioning satellites include:

****SURVEILLANCE satellites, the "spy" satellites used to observe activities of other countries.***

****WEATHER satellites, used in forecasting by photographing large storms and tracking motions of weather fronts within Earth's atmosphere.***

****REMOTE SENSING satellites, used in the study of resource deposits, geological variations in Earth's crust, and vegetation growth, and for detecting meteor craters.***

****COMMUNICATIONS satellites, used in transmitting radio signals for T.V. and telephone transmission.***

****NAVIGATIONAL satellites, used for precise determination of the locations of aircraft and ships.***

ACTIVITY

FIGURE 14-2. A several-minute exposure of the constellation Cepheus through which the satellite Echo I passed in 1972. (Photo courtesy of Vanderbilt Planetarium)

High-altitude satellites are too far away (and thus too faint) to spot with the naked eye or even with binoculars. But some satellites in lower orbits (300 miles or less) can easily be seen at night. Look for a tiny point of light that resembles a star moving slowly across the sky. At first you may think it's an airplane, but you'll soon notice the absence of sound or of flashing lights. As the satellite crosses the sky in about two to three minutes, you may even notice that it appears to twinkle, blink, or change in brightness erratically (Figure 14-2). This twinkling effect is caused by the satellite tumbling or rotating as it reflects sunlight down to Earth. About an hour after your sunset (an ideal time to look for satellites), the satellite itself can still be in sunlight because it is so high, thus making it visible.

Well after sunset, you may be lucky enough to spot a polar orbiting satellite. Such a satellite moves over the north pole, orbits Earth through changing latitudes, and then crosses Antarctica over the south pole. Because of its orbital path, a polar satellite is adept at mapping and observing the entire planet. To search for a polar orbiting satellite, first find the approximately north-south line. You can do this by facing the North Star (Polaris; see Activity 3), then turning to face south, which is behind your back. Scan the south for points of light, usually heading northward.

To enhance your view of satellites, use a pair of steadily held wide-field binoculars. For current information on passing satellites (and on the U.S. space shuttle or Russia's Mir space station), call **Sky & Telescope** magazine's skyline, (617) 497–4168 .

15 How Dark Is Your Backyard?

Understanding Light Pollution

Level: All

Objectives: To measure and to help reduce wasteful night lighting and its effects on the night sky

Materials:

- A night-sky observing site away from direct lighting glare
- Seasonal star charts (see Activities 4 through 7)
- A red-filtered flashlight (see Activity 2)

BACKGROUND

What is light pollution? You may be familiar with air, noise, and water pollution, but light pollution remains something of a mystery. Urban and suburban locations now bathe themselves in a vast pool of night lighting brought on by increasing population, technological growth, and competition for advertising space. Many homeowners misuse lighting by illuminating their property all night (Figure 15-1). Businesses add to dangerous glare (direct, blinding light) along roads with billboards and 24-hour energy-wasting signs. A shocking photograph of central North America (primarily the United States) at midnight from a 400-mile-high satellite reveals how wasteful night lighting has become (see Figure 15-2).

A century ago, the thought that night lighting could pose a threat to viewing the sky was virtually nonexistent. Today, many observatories, particularly in California, Arizona, and Hawaii, have had to seek legislation to halt the encroaching sea of light. Astronomers seek to understand the universe by observing extremely distant and faint objects, such as quasars and the remotest galaxies. Sadly, after having traveled many trillions of miles through space, the light from these objects is obliterated by bright cities during the last few thousand feet of its trip to Earth. As a result, astronomy research has become threatened by the perpetual brightening of the night sky by large cities (Figure 15-3).

Called sky-glow, the light that hurts our view of the night sky is caused largely by ineffi-

FIGURE 15-1. A house that is fully illuminated all night long is surely an expensive and bright place in which to live.

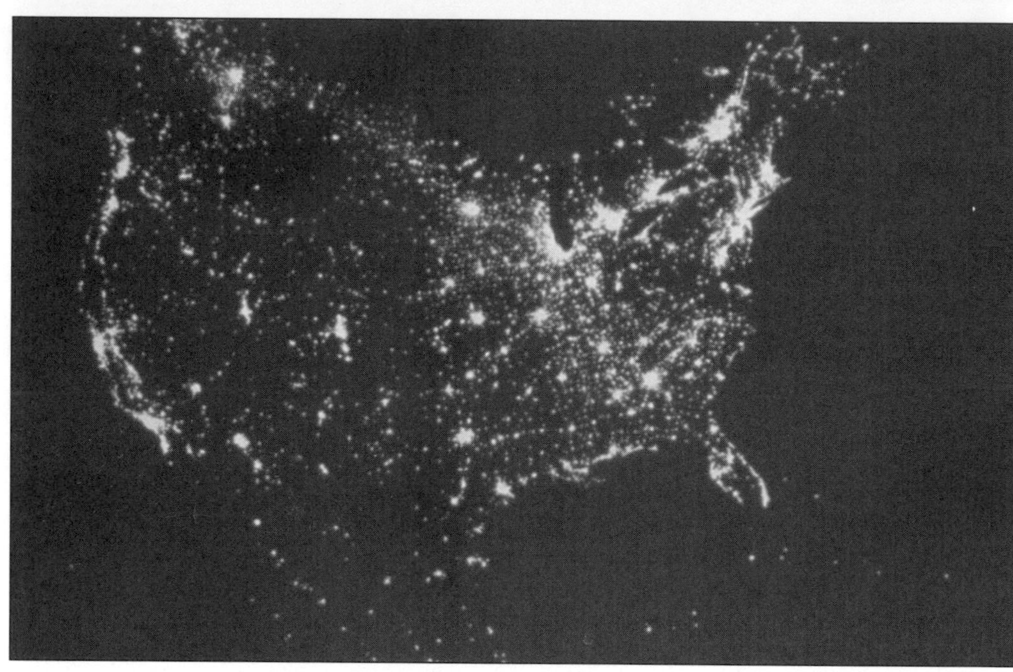

FIGURE 15-2. A 400-mile-high composite satellite photo of the United States at night reveals the shocking amount of stray light that is wasted in lighting up the night sky. (Photo courtesy of Dr. David L. Crawford, International Dark Sky Association)

cient, poorly placed street, house, and building lighting, much of it erected to allegedly deter would-be thieves. To date, no correlations between lighting and crime decrease have been made, though some studies have surprisingly shown an **increase** in crime with an increase in outdoor lighting. Such lighting usually comes from one of four principal types of lamps. **Incandescent lamps** are the least energy-efficient, though these are the most common consumer bulb. **Mercury vapor lamps** produce a glaring, bright blue-white hue. **High-pressure sodium lamps** emit a glaring peach-orange color. **Low-pressure sodium lamps** produce a yellowish glow and are the most cost-effective and energy-efficient of today's lights.

Astronomers are most eager to gain the public's support in using low-pressure sodium lamps, as these bulbs emit only one unchanging color (yellow) whose glow can be blocked using filters. Other types of light tend to emit all types of colors, making filtration difficult.

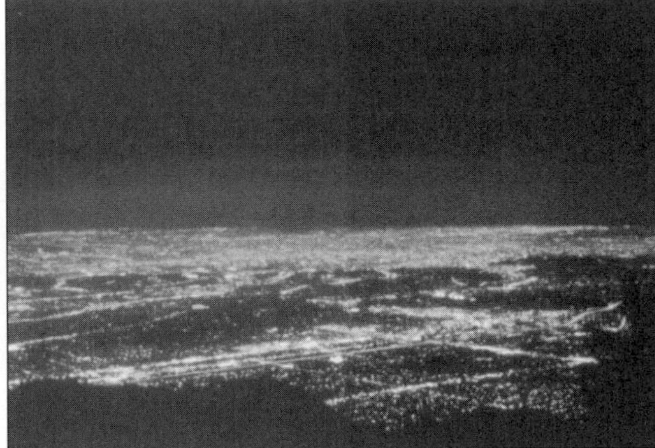

FIGURE 15-3. Cities continue to expand, grow, and fill the night sky with wasted light as shown in these comparison photos of Los Angeles in 1908 (left) vs. Los Angeles in 1988 (right).

Low-pressure sodium lamps can be mounted with boxes, shields, or covers to help reduce the amount of light spoiling the night sky (Figure 15-4). The International Dark-Sky Association (IDA), created by astronomer Dr. David Crawford, works to help citizens preserve the beauty of the night sky. For information on educating your community about poor lighting, contact IDA (listed in Appendix 2). With increased awareness of the light pollution problem, someday soon children will be able to see that the real Milky Way is not just a candy bar but an inspiring sight in the dark night sky.

FIGURE 15-4. An example of a properly designed light "box" fixture concentrating light only where it is needed—down, which eliminates direct glare and upward glow.

ACTIVITY

To determine the severity of light pollution in your celestial observing area, you can try to find various constellations, stars, and nebulae, using Table 15-1 as a guide to some objects to look for and the times of year for the search. Go outdoors on a clear, moonless night (remember, the moon adds light to the sky) and try observing objects listed. You can consult your star charts, using a red-filtered flashlight to help preserve your night vision. Once you've found an object you're looking for, check the appropriate line that describes the darkness rating of your site (superb, good, average, poor, or severely washed out).

So how bad (or good) is your observing site? Congratulations if you've checked off "superb" under each season; you are a most fortunate stargazer!

TABLE 15-1

This list of selected celestial objects can be used to judge the darkness of your observing site. With the naked eye, search for the object at least an hour after sunset during the specified season. (Descriptions of the North Star, Big Dipper, Little Dipper, and Cassiopeia can be used year-round, as they are circumpolar stars.)

IF:	YOUR LOCATION:	THEN the darkness rating of your sky is:
SPRING		
1) North Star (Polaris) invisible, Big Dipper barely visible	In a big city	Severely washed out
2) Big Dipper and stars Arcturus and Spica easy to spot	10–15 miles from city	Poor
3) Polaris, two brightest Little Dipper bowl stars, and most of Leo visible	15–50 miles from city	Average
4) Can see Dippers easily; can see star cluster Coma Berenices and Hercules	50–75 miles from city	Good
5) Can barely see the Hercules cluster; can see M13 as a fuzz spot	Remote farm, field, desert	Superb
SUMMER		
1) Summer Triangle barely visible overhead	In a big city	Severely washed out
2) Northern Cross (in Summer Triangle) barely visible	10–15 miles from city	Poor
3) Northern Cross easy to see, and stars of Lyra and Aquila (Southern Triangle) barely visible	15–50 miles from city	Average
4) Milky Way in Southern Triangle and Hercules somewhat visible	50–75 miles from city	Good
5) Milky Way shows much contrast, with view rich in binoculars	Remote farm, field, desert	Superb
FALL		
1) Summer Triangle and stars of Cassiopeia (W shape) barely visible	In a big city	Severely washed out
2) Cassiopeia easier to see and Pegasus barely visible	10–15 miles from city	Poor
3) Cassiopeia and Pegasus visible, and Andromeda barely visible	15–50 miles from city	Average
4) Pisces, and the Double Cluster between W and Perseus visible	50–75 miles from city	Good
5) Andromeda Galaxy (M31) barely seen and Milky Way easy to see in west	Remote farm, field, desert	Superb
WINTER		
1) Winter G (Orion et al.) barely visible	In a big city	Severely washed out
2) Pleiades cluster and V of Taurus barely visible, G easier to see	10–15 miles from city	Poor
3) Lepus (below Orion) and Orion's head and sword barely visible	15–50 miles from city	Average
4) Orion Nebula (M42) and six or seven of the Pleiades visible	50–75 miles from city	Good
5) Winter Milky Way over Orion, M44 cluster (in Gemini), and M42 easy to see	Remote farm, field, desert	Superb

SECTION II

The Moon

16 "Jack and Jill" and the Moon's Phases

Level: All

Objective: To learn the moon's phases

Materials:

- A clear view of successive lunar phases for about a two-week period after a new moon (see Appendix 3)
- Binoculars (optional)
- A bright-colored ball, such as a Ping Pong ball or softball
- A bright, unshaded lamp
- A darkened room
- A small colored sticker or small piece of tape
- Activity 17 (for reference)

From night to night, it is not difficult to notice the leisurely passing of the moon, closest of all celestial bodies to Earth, across the star-filled sky.

The moon's phases are the result of this movement, or revolution, around Earth and the moon's reflection of sunlight our way. As the moon orbits Earth, more and more of its surface faces direct sunlight **as seen by a viewer on Earth.**

To visualize this, create a model by letting your head represent Earth, a bright ball the moon, and a bright lamp the distant sun. Stand in the center of a darkened room and have a helper turn on the lamp, which should be placed several feet away. Next, put a small colored sticker or piece of tape on the ball as a landmark. Give the ball to your assistant and have him or her hold it directly between your head (Earth) and the lamp (the sun) with the sticker always facing you (the same side of the moon always faces Earth). This represents the moon's **new phase**, invisible to Earthbound viewers except during a solar eclipse (see Figure 16-1).

Now back to the real moon. Just after the appearance of the new moon, you can notice a sliver of light on the moon's left (or sun-facing) edge, representing the **young** or **waxing crescent phase** (Figure 16-2a; the moon's "age" refers to the number of days or hours since the last new phase). Notice that the "horns," or **cusps**, of the crescent always point away from the sun and that this phase sets shortly after the sun, so an unobstructed horizon is often required to see it best.

MOON LORE
Cultures around the world have created stories to explain the moon's mysterious changes. The Greeks personified the moon as Artemis, the Goddess of the Hunt, who provided evening light to assist hunters at only certain times during the month. In Central Africa the Masai explained the moon's changing appearance by telling that the sun and moon, once best friends, were beset by a violent quarrel during which the sun scratched the moon's face, giving it the scars seen today. Upon separating, the moon caught up with the sun every twenty-nine days, "hiding" its face in darkness. The sun, in embarrassment, turned a deep red and escaped below the horizon. In Greenland, the light of the perfectly round moon was an omen thought to cause young women to become pregnant!

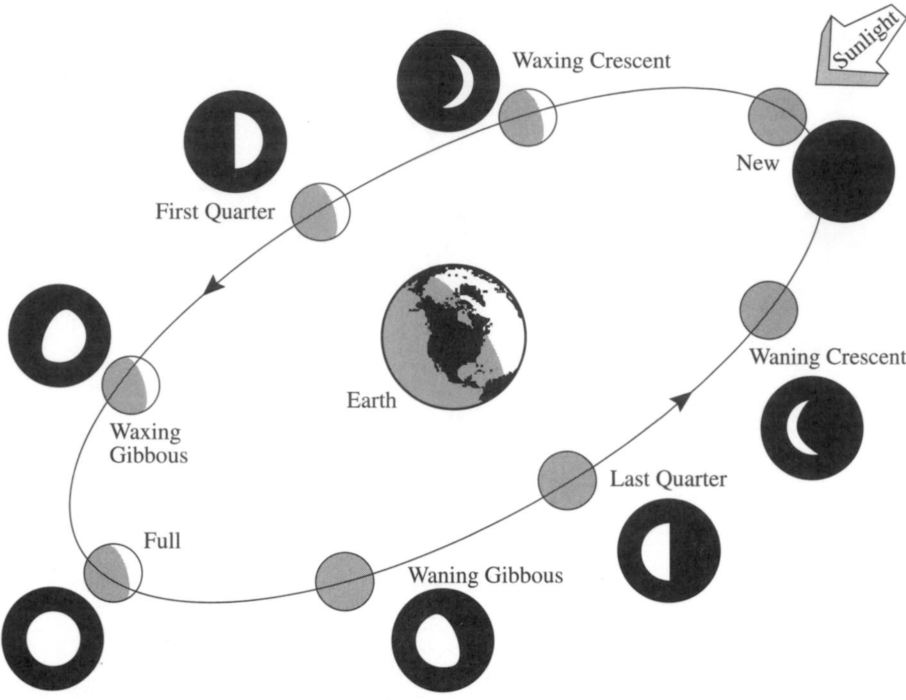

FIGURE 16-1. The lunar phases (in black circles) as seen with the naked eye and their locations in the moon's orbit.

FIGURE 16-2a. (left) Waxing crescent

FIGURE 16-2b. (right) First quarter

Back to the model. Next, have your assistant stop at the **first quarter phase** (Figure 16-2b), when the ball is one quarter its way around your head. The right half of the ball should be lit in the shape of a capital letter **D**, and the sticker should be on the line between the lit and unlit sides, called the **terminator.** Often called the half moon because it appears to be half full, this phase is correctly called the first quarter because the moon has now orbited one quarter (or 90°) of its way around Earth (see Figure 16-1). Typically rising around midday, when the sun is already halfway across the sky, the first quarter phase can be spotted by pointing your right hand toward the late afternoon sun and your left hand upward, perpendicular to your right hand (90°). Don't worry, it's perfectly normal to see the moon during the day!

FIGURE 16-2c. Waxing gibbous

FIGURE 16-2d. Full

FIGURE 16-2e. Waning gibbous

FIGURE 16-2f. Last quarter

FIGURE 16-2g. Waning crescent
(Photos courtesy of Lick Observatory)

As the moon waxes, or fills out, the first quarter phase gradually grows to a pear-shaped **gibbous** moon (Figure 16-2c). Thereafter, some fourteen days after new moon, the familiar **full** moon rises in the east as the sun sets, putting the moon 180° across the sky compared to the sun (Figure 16-2d). If you imitate this movement with your model, the ball should now be fully lit up.

After full, the moon's phase "fades" off, or wanes, first with the **waning gibbous phase** (Figure 16-2e), then with the **last quarter phase** (Figure 16-2f), which rises around midnight and sets the following day around noon. Now the ball in your model will have moved (orbited) three quarters of its way around your head. The left portion should now be lit, with the sticker again on the terminator. Finally, just after the **waning crescent phase** (Figure 16-2g), the monthly lunar cycle concludes when the moon returns to the **new phase** (directly between your head and the lamp), 29 ½ days after the previous new phase.

Everyone knows the nursery rhyme "Jack and Jill," but did you know that this rhyme tells a story about the moon's changing phases over the course of a month?

FIGURE 16-3a. As the moon phases, you may imagine the story of Jack and Jill among the dark lunar seas. Compare these same phases with Figure 16-2 and Activity 17 to identify the specific maria. (Photos courtesy of Vanderbilt Planetarium)

As you watch the moon's phases, try carefully observing the grayish patterns on its surface, called **maria**. Using the naked eye or binoculars, can you visualize the story of Jack and Jill using the maria? The rhyme comes from ancient Scandinavia. **Jakka**, the Scandinavian word for "increase," represents the waxing lunar phases (the phases after new but before full). **Bila**, their word for dissolve or decrease, represents Jill, or the waning phases after full and before new moon. At first quarter an outline of a head, arms, body, and legs (Jack) may be perceived. At the full phase you may envision Jill standing to the left of Jack, while at last quarter, "Jack falls down" and Jill is last to fade just before new moon (compare Figure 16-2a-f with Figure 16-3a-g).

Using your model, you can also illustrate how and when lunar and solar eclipses occur. At the new phase, have your assistant block the lamp from your view with the ball. For others in the room, the lamp will still shine, but for you (Earth) the lamp will have appeared to

FIGURE 16-3b.

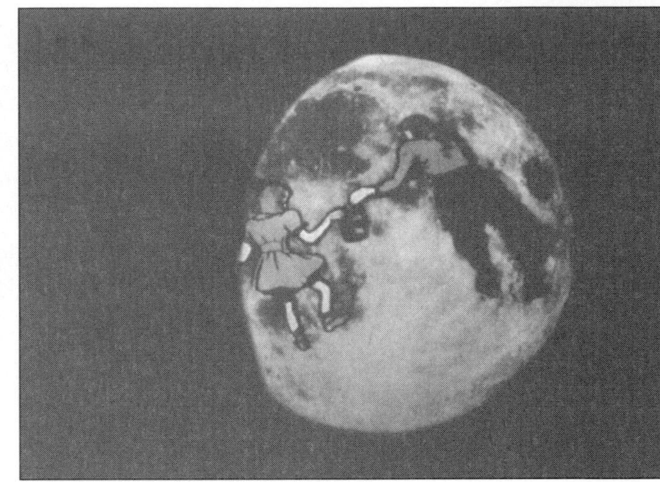

FIGURE 16-3c.

FIGURE 16-3d.

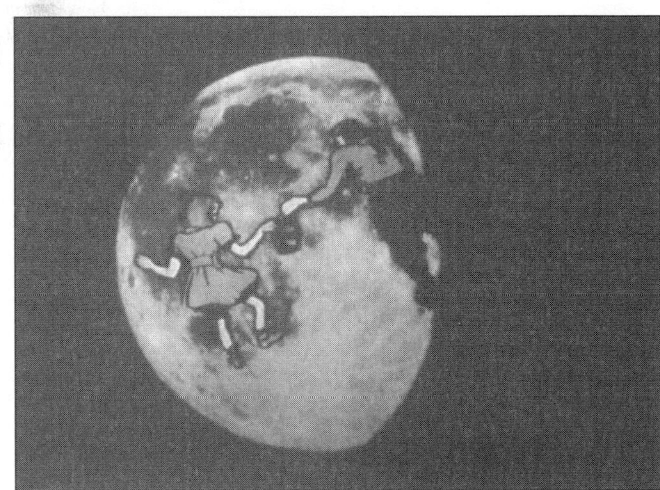

FIGURE 16-3e.

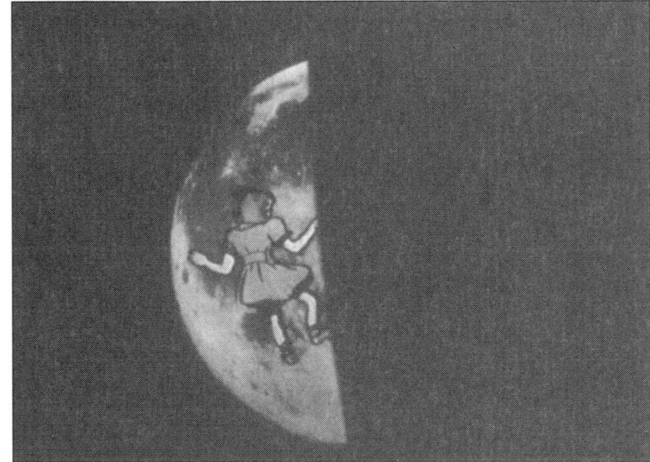

FIGURE 16-3f.

FIGURE 16-3g.

disappear behind the ball (moon), creating a model solar eclipse. Next have your assistant move to the full phase without letting lamplight fall on the moon. The ball will be in your head's shadow, simulating a lunar eclipse. During an actual lunar eclipse, observers on Earth's night side see the moon pass through Earth's shadow, turning a variety of interesting hues as a result of sunlight filtered through Earth's atmosphere.

Once you have observed the moon both during the day and night, you will become more accustomed to how it orbits Earth and why it appears as it does. Only then may you begin to understand how the ancients saw such importance in the lunar phases.

17 Oceans on the Moon?

Looking for Lunar "Seas"

Level: Intermediate and advanced

Objective: To identify the dark plains on the moon

Materials:

- A clear night with the moon visible
- Binoculars or a small telescope
- A red-filtered flashlight (see Activity 2)
- Activity 16 (for reference)

BACKGROUND

Look at the moon on a clear night and you'll notice distinct light and dark regions on its surface. Ancient civilizations explained the moon's presence and motion with stories about creatures they imagined among the patterns formed by these light and dark markings. The lighter areas are the **lunar highlands**—usually mountainous terrain covered by impact craters. The darker areas are the **lunar lowlands**, also called the **lunar maria**. "Maria" is the plural of "mare," the Latin word for "sea." The lunar maria are actually dry land areas, but for centuries it was thought that they were oceans of water teeming with life.

In his 1609 work **Siderius Nuncius** (The Starry Messenger), Italian astronomer Galileo Galilei saw maria as "large ancient spots on the moon . . . which appear to be even and uniform," referring to their lack of craters. Galileo also thought the maria resembled what Earth's oceans might look like from space. Maria are really large depressions called **crater basins**. Some three to four billion years ago, meteor collisions not only scarred the moon's outer crust but also were occasionally energetic enough to impale the mantle, just below the crust. Lava oozed upward at the impact site, filled the crater basin, and cooled into the smooth plains that we see today.

ACTIVITY

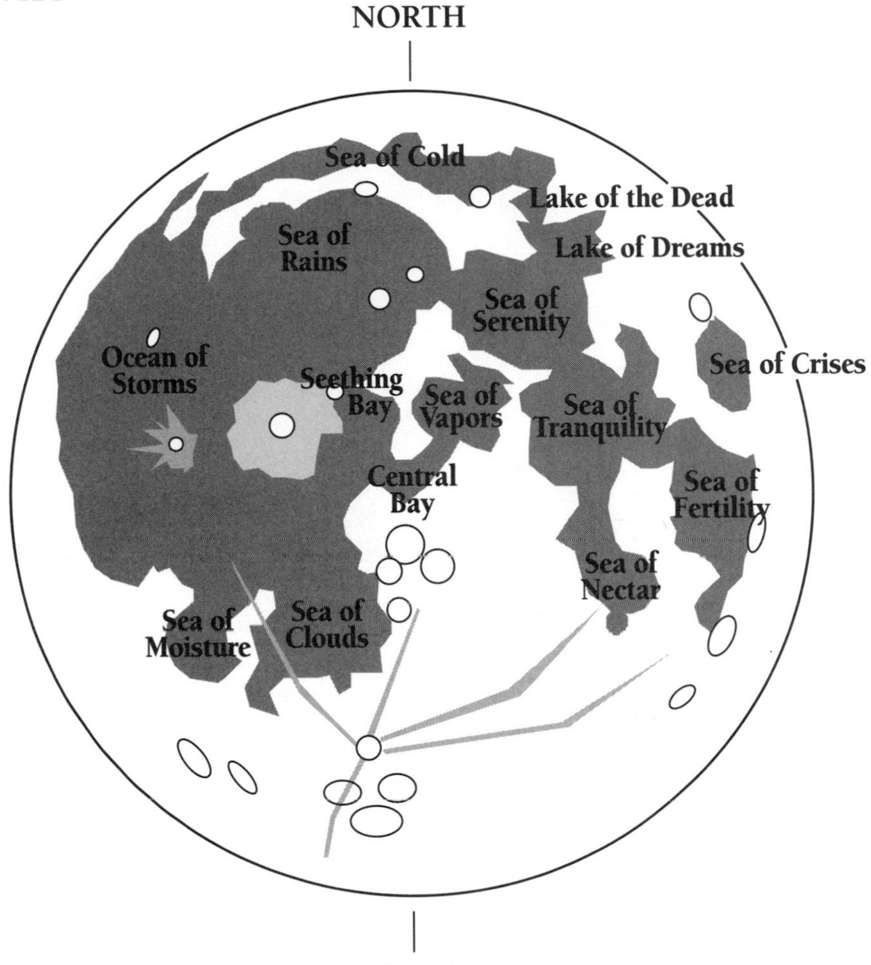

FIGURE 17-1. **Here, the entire Earth-facing side of the moon is shown, as are its major maria (lunar seas).**

Astronomers have identified and named the major lunar maria, just as people have given names to geographical features here on Earth.

Try to identify the maria described below. Note cratering and texture differences between the maria and the lunar highlands. You can use your naked eye, binoculars, or a small telescope. (Refer to Figure 17-1.)

Young Crescent Moon

Observing the two-day-old waxing crescent moon following sunset, you should see at least two major maria. (The "age" of the moon refers to the number of days or hours since the last new phase; check phases in Appendix 3 or Activity 16.) The **Sea of Crises**, a 270-mile by 350-mile oval basin, was the site of the Soviet Union's Luna 15 landing in 1969. Luna 23 in 1975 and Luna 24 in 1976 also landed here, with Luna 24 returning soil samples from the Sea of Crises.

Below the Sea of Crises, locate the **Sea of Fertility**—the landing site of the 1970 Luna 16 mission, the first Soviet voyage to return moon rocks and soil to Earth. The Sea of Fertility is part of the imagined figures of Jack and Jill that are created by the lunar maria. The Sea of Fertility is one of Jack's legs as described in Activity 16. We'll describe the rest of these figures as we identify more lunar maria in this activity.

First Quarter Moon

The first quarter moon, seen about a week after new moon, is best remembered as looking like a capital "D." Locate the **Sea of Serenity** (Jack's head), which has nearly the same area as the state of Nevada. With a 6- or 8-inch reflecting telescope, you can see this area's mountainous boundary encompassing the wavelike ripples and tiny craters of its plain. Just above the Sea of Serenity you'll see two wispy maria, the **Lake of the Dead** and the **Lake of Dreams**. The **Sea of Cold** is a small, faint maria above the Lake of the Dead.

Below the Sea of Serenity is the **Sea of Tranquillity** (Jack's body), measuring about 400 by 550 miles (as large as Texas). The Sea of Tranquillity was the site of the first manned lunar landing—on July 20, 1969, by Apollo 11 astronauts Edwin Aldrin and Neil Armstrong. Just below the Sea of Tranquillity, locate the **Sea of Nectar** (Jack's other leg), which is about 240 miles across.

Waxing Gibbous Moon to Full Moon

The gibbous (oval-shaped) phase occurs between first quarter and full (when the moon is **waxing** or filling out) and between full and last quarter (when the moon is **waning** or diminishing). Look for the "pail" that Jack and Jill are holding, formed by the **Sea of Vapors**. Jill's arm is formed by **Seething Bay** and **Central Bay**. Jill's head—the **Sea of Rains**—begins to appear about five days before full moon. One of the moon's largest maria, this sea is nearly the size of Alaska and has been visited by the Soviet craft Luna 17. Jill's legs, composed of the **Sea of Clouds** and the **Sea of Moisture**, are less distinct than Jack's.

Just before full moon, look for the **Ocean of Storms**, the largest of the lunar basins. Nearly twice the size of Alaska, the Ocean of Storms has been visited by space vehicles from the United States (Surveyors 1 and 3 and Apollo 12) and from the Soviet Union (Luna 5, 7, 8, 9, and 13).

These are some of the major maria visible on the side of the moon that always faces Earth. Other lunar maria are on the unseen side of the moon, invisible to Earthbound observers. The Apollo space program produced photos of the moon's far side, resulting in maps as detailed as those of the side we see from Earth. Can you find other maria that we have not listed (see *Exploring the Moon* by Ernest Cherrington, listed in Appendix 1).

18 Potholes on the Moon?

Looking for Lunar Craters

Level: All

Objectives:

- To find specific craters on the moon
- To observe various lunar craters and note their similarities and differences

Materials:

- Binoculars or a telescope
- Moon map (Figure 18-1)
- Photocopies of Moon
- Observation Form (included with this activity)
- Pencil
- Clipboard

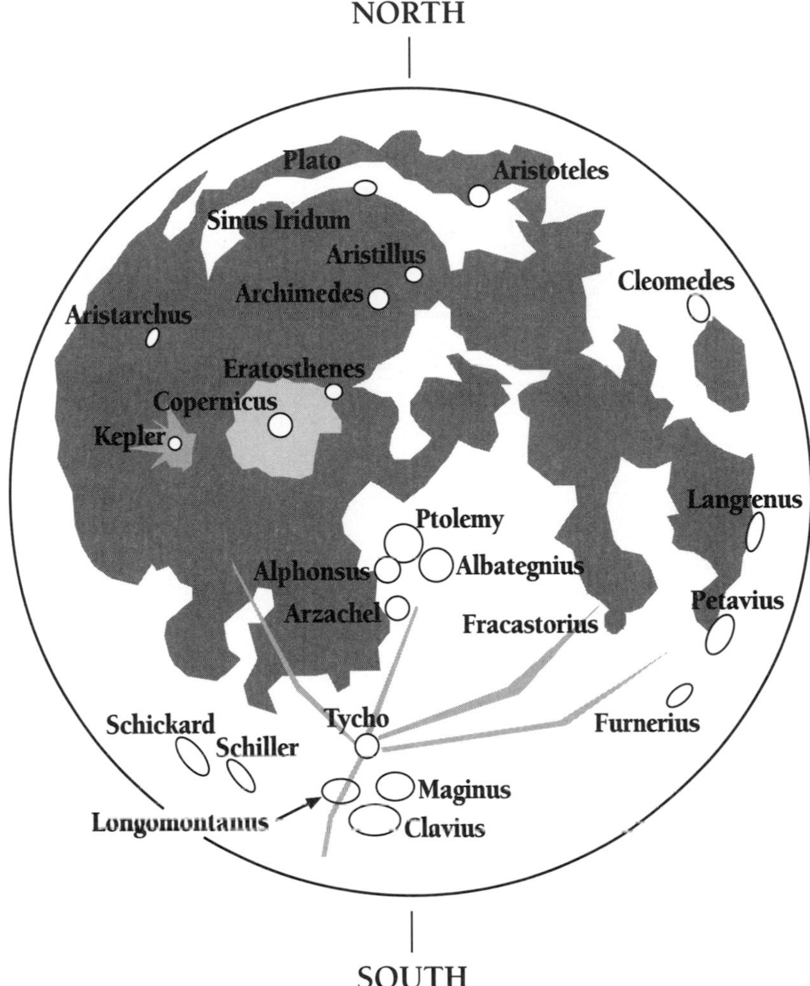

FIGURE 18-1. The moon's most prominent craters.

BACKGROUND

Lying nearly a quarter of a million miles from Earth, the moon presents a magnificently rugged and varied terrain for earthbound lunar explorers to investigate. The last activity explored the dark lunar seas, or maria, that dominate the view. Now we shall observe some of the myriad potholes, called **craters**, that are scattered across the scarred lunar surface. Craters come in a wide variety of sizes, as shown in Figure 18-2.

Scientists believe that most of our moon's craters, like those found on other moons and planets throughout the solar system, were caused by collisions with debris left over from the formation of the solar system some 4.5 billion years ago. Although most of Earth's meteor craters have since been either eroded away or overgrown with vegetation, the airless, waterless moon stands as an eternal record of our own world's violent beginnings.

ACTIVITY

Viewing and identifying craters on the moon for the first time can be likened to your first visit to a big city. Where would you go? There are so many things to see, it's hard to know just where to begin. You might go to a famous museum, various parks and monuments, and a planetarium or science museum. But selecting just a handful of stops would only skim over what the city has to offer.

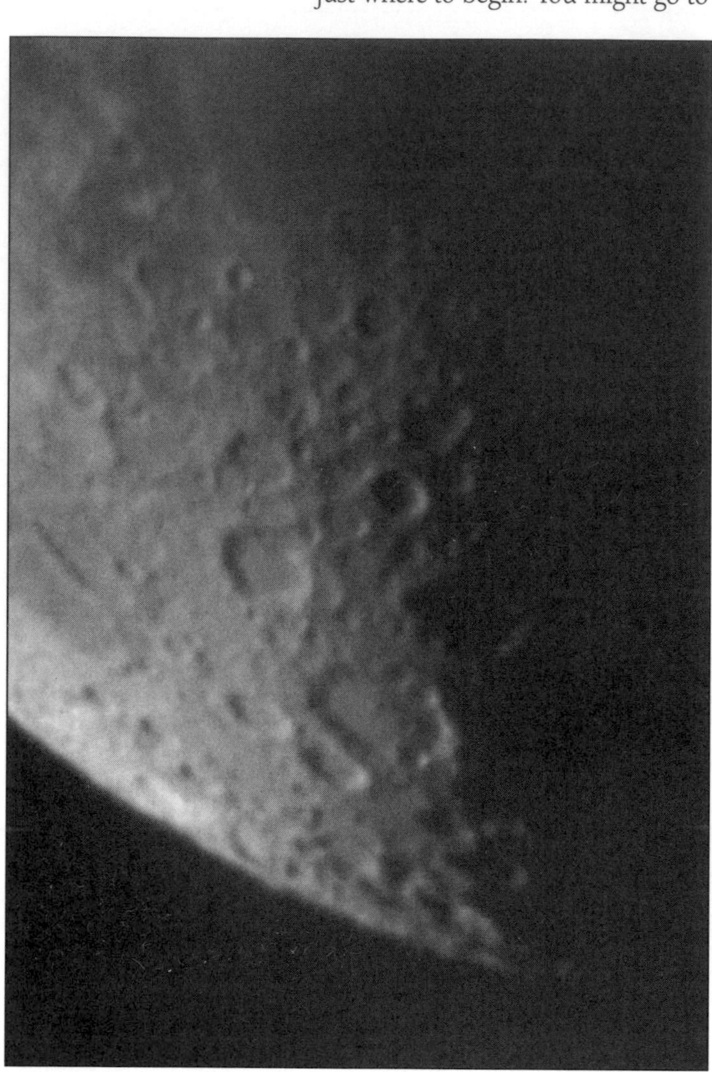

FIGURE 18-2. The moon is littered with craters. The largest crater in view here is Clavius, found near the moon's south pole.

The same is true of the moon. In this activity, we shall just **begin** to explore the myriad craters that scar our closest space neighbor. It would be impossible for a book such as this to list and describe every crater visible through small backyard telescopes. There are just too many of them! Therefore, the following discussion and map (Figure 18-1) are limited to the finest, most prominent craters.

As noted in the previous activity, different moon phases are better for lunar sightseeing than others. Little or no detail may be spotted for the first couple of nights after the new phase, but as the moon progresses night after night toward the east, more and more craters become visible. The best time to view a specific crater is when it is on or near the moon's **terminator**, the day/night line that runs from the moon's north pole to its south pole. At these times, the sun's light striking the crater gives the greatest three-dimensional relief. If sunlight just strikes a crater's rim, the crater will look like a bright, bottomless ring. As the sun's elevation increases, its light travels down the steep, clifflike walls until it floods the crater's floor. Continue to watch a particular crater for several nights to see what effect the sun's changing angle has on its appearance.

Try sketching as many of these craters as possible on a copy of the Moon Observation Form. Zero in on one or two of the craters and sketch what you see. A soft pencil with a blunt point will give the best results. Begin by lightly drawing in the outline of the crater. When you are satisfied with the shape, begin to shade in the dark areas with the pencil. Be careful not to shade the bright areas. Finally, smudge the lead on the paper with your finger, eraser, or other blunt object to create the proper blend of tones in your drawing. Most importantly, don't rush. Take your time, and soon you will have a nice collection of moonscapes. If you have access to a 35mm single-lens reflex camera, it is also possible to photograph the moon. For a complete discussion, see Activity 50.

Following is a short visitor's guide to some of our moon's finest craters, including the best lunar phases for viewing. For a depiction of the various lunar phases, see Figure 16-2 (in Activity 16); for the dates of lunar phases, see Appendix 3.

Moon Observation Form

Date: ______________________________ Time: ______________________

Telescope: ___________________________ Power: _____________________

Crater(s): ____

Notes: ____

____

____

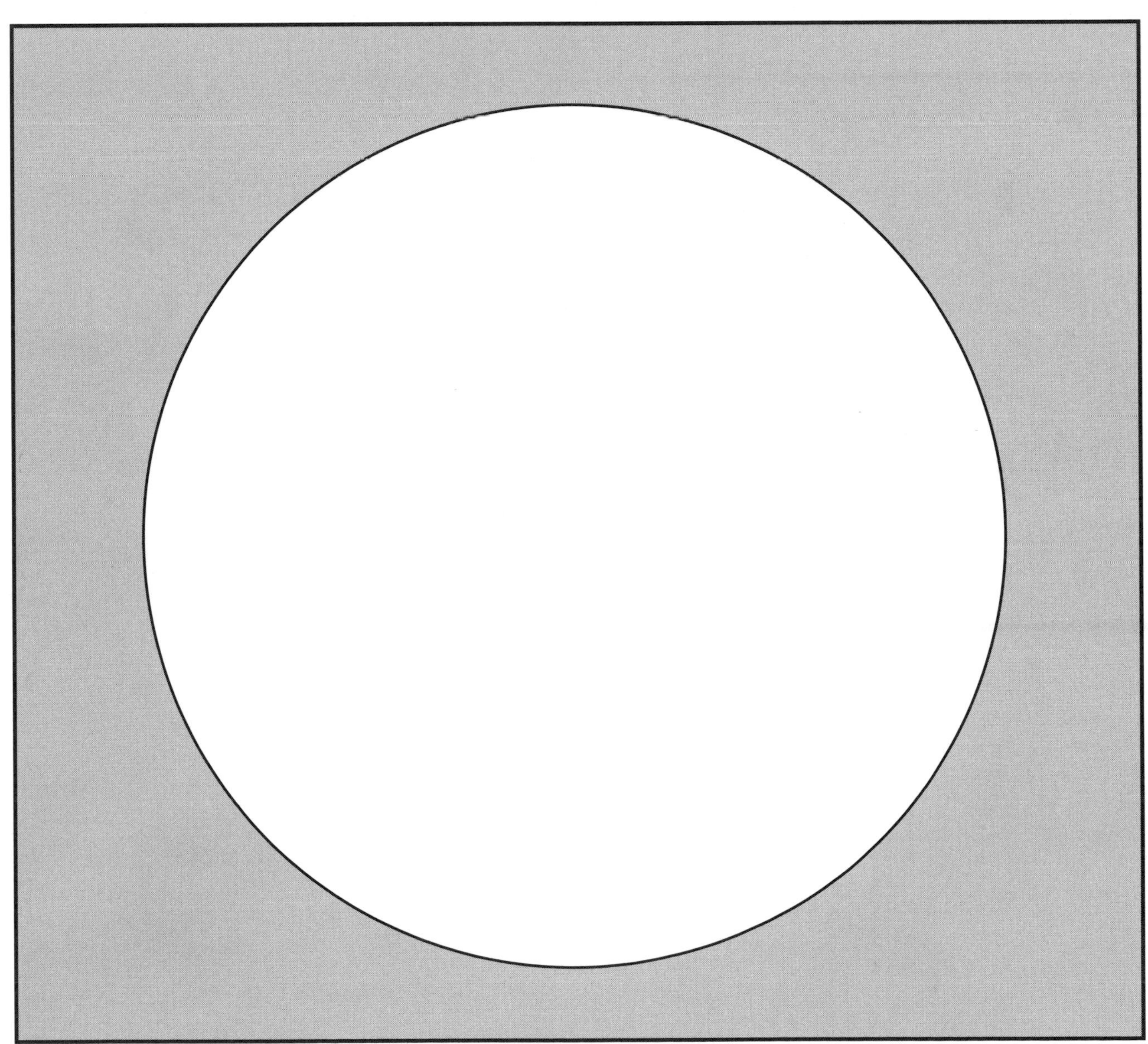

Albategnius: 81 miles in diameter, 14,000-foot-high walls. Look for an off-center peak rising from the dark crater floor. (Best phases for viewing: first quarter and last quarter)

Alphonsus: 64 miles by 73 miles across. Alphonsus has been observed to have, on extremely rare occasions, a cloud of vaporous emissions floating above it. (Best phase for viewing: waxing gibbous and waning crescent)

Archimedes: 51 miles in diameter. A portion of its walls were flooded by lava during the moon's volcanic era. (Best phases for viewing: waxing gibbous and waning crescent)

Aristarchus: 25 miles in diameter. Highlighted by a magnificent system of rays spraying out from its rim into the surrounding Ocean of Storms. (Best phases for viewing: waxing gibbous and waning crescent)

Aristillus: 36 miles in diameter. Aristillus is also highlighted by bright rays, making the crater easy to find. (Best phases for viewing: waxing gibbous and waning crescent)

Aristoteles: 55 miles in diameter, 12,000-foot-high walls. The steeply banked walls plummet to a smooth, bright floor. (Best phases for viewing: waxing crescent and waning gibbous)

Arzachel: 59 miles in diameter, 13,000-foot-high walls. (Best phases for viewing: waxing gibbous and waning crescent)

Clavius: 132 miles by 152 miles across. Clavius is the second-largest crater on the moon's earthbound side. On the floor of Clavius are several smaller impact craters, with 30-mile-wide Clavius B the most obvious. (Best phases for viewing: waxing gibbous and waning crescent)

Cleomedes: 81 miles by 92 miles across, 15,000-foot-high walls. Highlighted by a rough floor. (Best phases for viewing: waxing crescent and waning gibbous)

Copernicus: 60 miles in diameter. Copernicus is marked by a brilliant ray system that bursts into view against the darker background of the Ocean of Storms. Also watch for its central mountain peak. (Best phases for viewing: waxing gibbous and waning crescent)

Eratosthenes: 37 miles in diameter, 13,000-foot-high walls. (Best phases for viewing: waxing gibbous and waning crescent)

Fracastorius: 73 miles in diameter. Partially destroyed crater ring caused when part of its north wall was washed away when a flood of hot lava from the neighboring maria filled its floor. (Best phases for viewing: waxing crescent and waning gibbous)

Furnerius: 81 miles in diameter. Found along the moon's edge (called the **limb**). (Best phases for viewing: waxing crescent and waning gibbous)

Kepler: 20 miles in diameter. Though small, Kepler is one of the most prominent craters on the moon thanks to its bright ray pattern. Kepler reminds many observers of a miniature Copernicus, which, incidentally, is just to the east. (Best phases for viewing: waxing gibbous and waning crescent)

Langrenus: 85 miles in diameter. An amazing crater to watch from waxing crescent to full moon. As the sun rises higher in the sky above it, Langrenus seems to almost catch fire as its floor is transformed from a dull gray to a brilliant white. (Best phases for viewing: waning crescent through full moon)

Longomontanus: 107 miles in diameter. (Best phases for viewing: waxing gibbous and waning crescent)

Maginus: 100 miles by 105 miles across. (Best phases for viewing: waxing gibbous and waning crescent)

Petavius: 99 miles by 110 miles across. Watch for its 8,200-foot-high central peak rising above the crater floor. (Best phases for viewing: waxing crescent and waning gibbous)

Plato: 64 miles by 67 miles across. Its floor is the darkest of any crater found on the moon's earth-facing side. (Best phases for viewing: waxing gibbous and waning crescent)

Ptolemy: 93 miles in diameter, 10,000-foot-high walls. (Best phases for viewing: waxing gibbous and waning crescent)

Schickard: 135 miles by 150 miles across. Take a look at its unusual floor, highlighted by two darker regions sandwiching a brighter central plateau. (Best phases for viewing: waxing gibbous and waning crescent)

Schiller: 48 miles by 113 miles across. Looks like a giant footprint. (Best phases for viewing: waxing gibbous and waning crescent)

Sinus Iridum: 160 miles in diameter. Originally a complete crater, the southern wall was totally washed away when lava from Mare Imbrium crashed against it. (Best phases for viewing: waxing gibbous and waning crescent)

Tycho: 56 miles in diameter. The most spectacular crater of all during the gibbous and full phases, with dazzling rays scattering radially outward from the crater for hundreds of miles. (Best phases for viewing: waxing gibbous through waning gibbous)

These are just a few of the hundreds of craters that can be seen through small telescopes. If you enjoy locating and identifying moon craters and would like to spot even more, look for the book *Exploring the Moon through Binoculars and Small Telescopes* by Ernest Cherrington. Another excellent book for lunar observers is Michael Kitt's *The Moon: An Observing Guide for Backyard Telescopes*. Both are listed in Appendix 1.

19 Where Did the Astronauts Land on the Moon?

Level: Intermediate and advanced

Objective: To locate the landing sites of the Apollo missions

Materials:

- A telescope
- Moon map (Figure 19-2)

BACKGROUND

"That's one small step for a man, one giant leap for mankind." With those words, spoken July 20, 1969, Apollo 11 astronaut Neil Armstrong became the first human to walk on the moon. Never before had humanity become so entranced by a scientific event. Around the world, people stopped whatever they were doing to watch as Armstrong and Edwin Aldrin set up scientific instruments and the American flag and gathered samples of the moon's soil and rocks. Michael Collins, Apollo 11's third astronaut, shared the world's amazement at the feat as he orbited the moon high overhead.

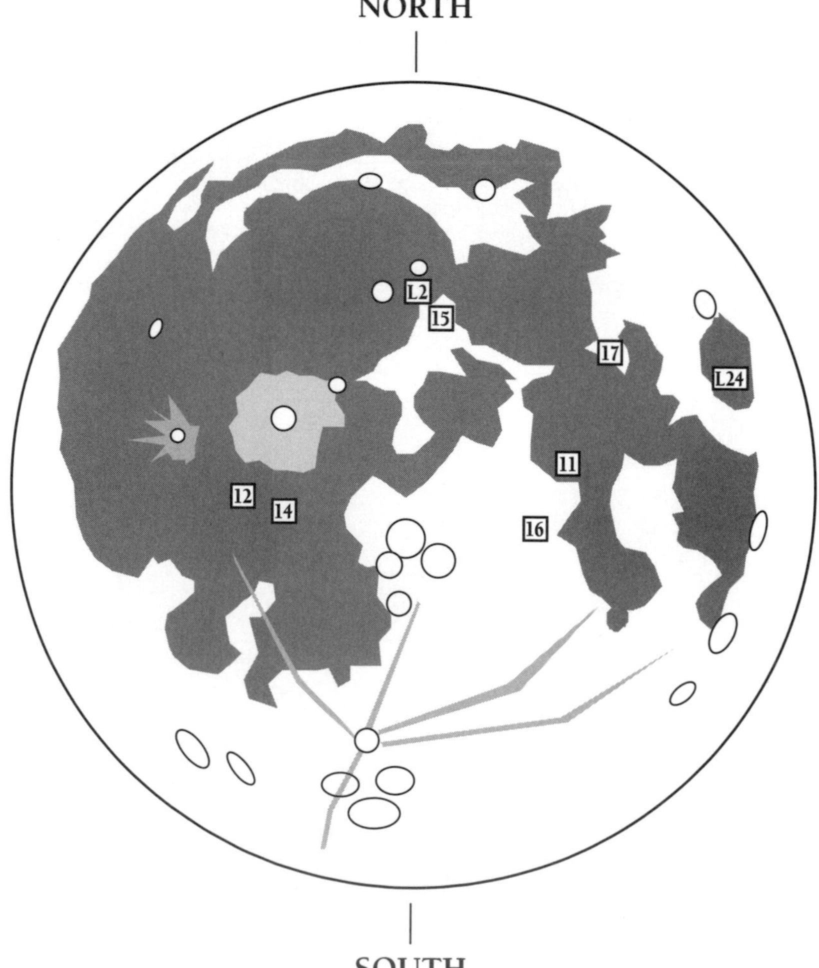

Figure 19-1. Locations of the Apollo landing sites as well as the landing sites of the unmanned Russian missions Lunokhod 2 and Luna 24.

The steps of Armstrong and Aldrin were the culmination of the Apollo space program, an effort dedicated to landing a man on the moon and returning him safely to Earth, as President John F. Kennedy explained it in 1963. Six years later, the program, directed by the National Aeronautics and Space Administration (NASA), capped its effort with the mission of Apollo 11.

Apollo 11 was the first of six Apollo missions to land on the moon. All took place from July 1969 to December 1972 (Figure 19-1). Those were the golden years of America's space program, years that generated great excitement and great curiosity about our nearest neighbor in space. To try to rekindle that excitement, this activity lets you "visit" each of the Apollo landing sites without ever leaving your backyard.

ACTIVITY

The accompanying map of the moon (Figure 19-2) shows the landing sites of the six Apollo moon flights (and of two unmanned Soviet craft). Each Apollo landing is labeled with its corresponding mission number (11 for Apollo 11, etc.). Though the map shows the moon at its full phase, it is recommended that you view the landing sites at other phases. Recall from earlier discussions that, while there is absolutely no danger to your eyes in looking at the full moon, most people find its brightness uncomfortable during that phase. Therefore, it is recommended that the moon be observed during its crescent, quarter, or gibbous phases.

View each Apollo site with a telescope, using at least 75-power to 100-power. This will permit close-up examination of the areas. Survey each site for craters and other features. Imagine what it must have been like to see those same features not from 240,000 miles away (as we are here on Earth) but instead from only a mile or less above the surface. As you moon-hop from one Apollo site to another, think of the following short summaries of each mission. In each case, only the commander and lunar-module pilot actually landed and walked on the moon; the command-module pilot stayed behind in lunar orbit.

Figure 19-2. A photograph from the Apollo 16 mission. (NASA photo)

APOLLO 11

Dates: July 16–24, 1969

Landing site: Sea of Tranquillity

Astronauts: Neil Armstrong (commander) Edwin Aldrin (lunar-module pilot) Michael Collins (command-module pilot)

Accomplishments: First two astronauts to walk on the moon; planted American flag; conducted limited scientific experiments; left plaque that stated "We came in peace for all mankind"; returned lunar soil samples.

APOLLO 12

Dates: November 14–24, 1969

Landing site: Ocean of Storms

Astronauts: Charles Conrad (commander) Alan Bean (lunar-module pilot) Richard Gordon (command-module pilot)

Accomplishments: Landed within walking distance of Surveyor III, an unmanned lunar probe that landed in April 1967; returned parts of the Surveyor spacecraft to study effects of the moon's harsh environment on metals.

APOLLO 13

Dates: April 11–17, 1970

Landing site: Scheduled to land at Fra Mauro; mission aborted in mid-flight

Astronauts: James Lovell (commander) Fred Haise (lunar-module pilot) Jack Swigert (command-module pilot)

Accomplishments: When explosion of oxygen tank en route to the moon forced cancellation of landing mission, astronauts circled the moon and returned safely to Earth.

APOLLO 14

Dates: January 31–February 9, 1971

Landing site: Fra Mauro

Astronauts: Alan Shepard (commander) Edgar Mitchell (lunar-module pilot) Stuart Roosa (command-module pilot)

Accomplishments: Returned lunar soil samples; Shepard hit golf balls with an improvised golf club.

APOLLO 15

Dates: July 26–August 7, 1971

Landing site: Hadley Rill, near the Apennine Mountains

Astronauts: David Scott (commander) James Irwin (lunar-module pilot) Alfred Worden (command-module pilot)

Accomplishments: First mission to bring along the lunar rover "moon buggy," enabling astronauts to travel farther from landing site; found 4.5-billion-year-old "Genesis rock" that confirmed scientists' estimate of the age of the solar system.

APOLLO 16

Dates: April 16–27, 1972

Landing site: The crater Descartes

Astronauts: John Young (commander) Charles Duke (lunar-module pilot) T. Kenneth Mattingly (command-module pilot)

Accomplishments: Using lunar rover, gathered soil samples that proved the moon was once volcanically active.

APOLLO 17

Dates: December 7–19, 1972

Landing site: Adjacent to the craters Taurus and Littrow

Astronauts: Eugene Cernan (commander) Harrison Schmidt (lunar-module pilot) Ronald Evans (command-module pilot)

Accomplishments: Using lunar rover, explored northeastern quadrant of the moon; Schmidt became the first scientist to visit moon; left plaque on the moon that read "Here man completed his first explorations of the moon. May the spirit of peace in which we came be reflected in the lives of all mankind."

The Apollo program signaled the end of the race that had pitted the United States against the Soviet Union to see who could first land astronauts on the moon. Though the Soviets never did send any cosmonauts to the moon, they did send unmanned missions. After Apollo 17, two Soviet unmanned spacecraft completed missions to the moon. Lunokhod 2 (shown on the map as L2) landed in January 1973 and was a robotic spacecraft capable of crawling across a small portion of the moon's surface. Luna 24 (labeled L24 on the map) landed in August 1976 in the Sea of Crises and brought back soil samples to scientists on Earth.

Table 19-1 lists the best phases for finding each of the Apollo landing sites.

TABLE 19-1

Best Phases for Finding the Apollo Landing Sites

MISSION	SITE	PHASE
Apollo 11	Sea of Tranquillity	Waxing crescent to first quarter
Apollo 12	Ocean of Storms	Waxing gibbous
Apollo 14	Fra Mauro	Waxing gibbous
Apollo 15	Hadley Rill	First quarter
Apollo 16	Descartes	Waxing crescent to first quarter
Apollo 17	Taurus/Littrow	Waxing crescent
Lunokhod 2	Sea of Rains	First quarter to waxing gibbous
Luna 24	Sea of Crises	Waxing crescent to first quarter

Each night, as you look up at the moon, try to imagine what it must have been like to travel there. Imagine going to a place where there is no air to breathe and no water to drink;

where temperatures in the daytime average 250°F above zero, while at night they plummet to –200°F. The bravery of the Apollo astronauts cannot be overstated. And while no missions to the moon will take place in the foreseeable future, we hope that one day humans will again walk on its surface, resurrecting the magic and awe of the Apollo missions.

20 Observing the Odd Lunar Crescents of Summer and Winter

Level: All

Objective: To observe lunar "nighttime" and tilting lunar crescents

Materials:

- A bright ball, such as a Ping Pong ball or softball
- A lamp in a darkened room
- A small hand-held mirror
- A clear view of the crescent moon just after sunset (See Activity 16 for an explanation of the lunar phases; also see the lunar phase table in Appendix 3.)

BACKGROUND

The crescent moon, visible after sunset or before sunrise as a thin sliver of light near the horizon, has some odd features that many observers casually overlook. As you watch the sky begin to darken after sunset, watch the young crescent moon carefully and you will notice the entire outline of the moon, not just the crescent (Figure 20-1). Why should this be? Similarly, throughout the year you may notice that the crescent moon of summer looks just a little different from that of winter, as seen in Figures 20-2a and 20-2b. Let's investigate these lunar mysteries a bit more.

How is it possible to trace the moon's entire outline when it is only partially sunlit? The brighter, white portion of the moon is the region in direct sunlight, essentially "daytime" on the moon. Notice that this daylight gradually fades into a dark gray-black region of the moon along a line called the terminator. Low-powered binoculars or a camera viewfinder will help you see this dark region, called lunar "nighttime," more clearly. It is the portion of the moon that is not in direct sunlight. So why is it visible?

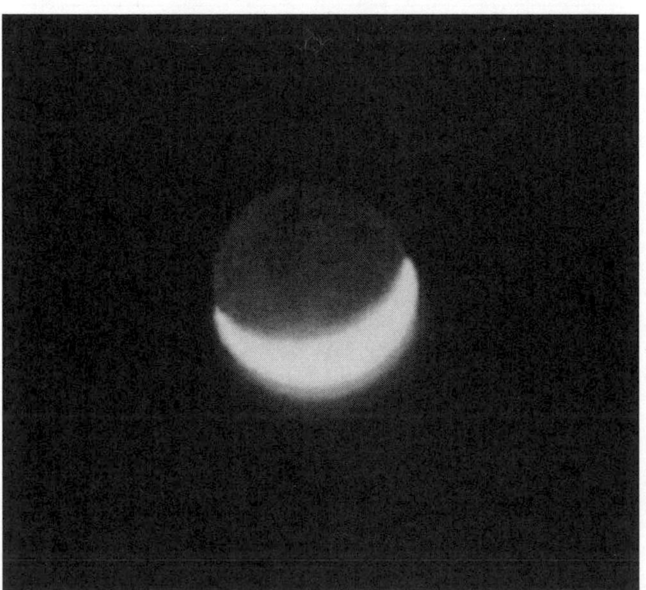

FIGURE 20-1. As shown, a slightly overexposed photo of a crescent moon reveals the illuminated night side, the result of reflected sunlight off Earth, called Earthshine.

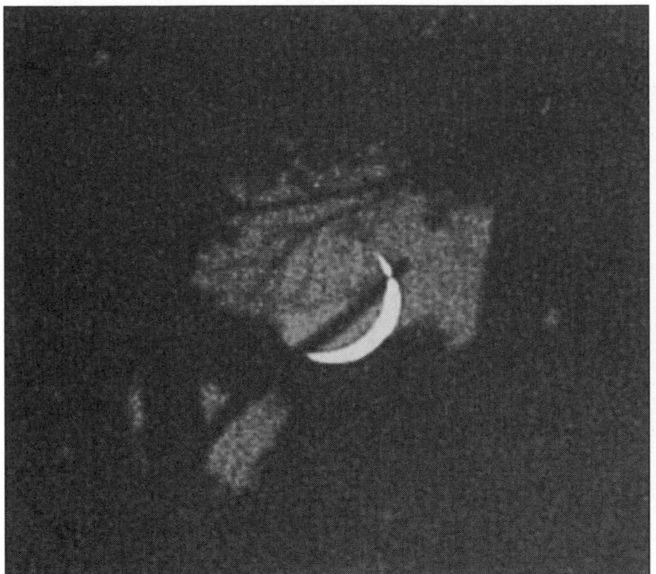

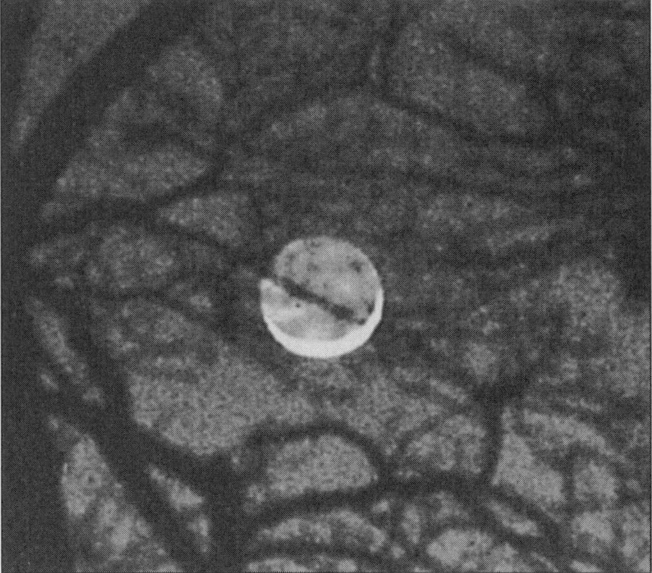

FIGURE 20 2a. (above left) The summer waxing crescent moon, shown here fifty-six hours after new moon. Note the tilt of the crescent as compared to FIGURE 20-2b. (above right) The winter waxing crescent moon, shown here thirty-six hours after new moon. Here, the cusps are tilted upward.

ACTIVITY

To understand this effect a little more clearly, create a model in which your head is the planet Earth, a ball is the moon, and a lamp represents the sun. With the lamp turned on in an otherwise completely darkened room, walk several feet from the lamp. As you do, hold the ball in front of you so that it reflects the lamplight but does not completely block the lamp. Hold the small mirror near your head facing the ball to reflect light toward the ball. Now, look carefully at the **direct** light hitting the ball (the brightest part) and the **indirect** light hitting the ball (the dimmer portion, in shadow). What lights up this side of the ball?

As the lamp shines, light from it also reaches the walls, floors, and ceiling of the room. These surfaces, including the surface of your body, then act like mirrors, letting light bounce, or reflect, off them. Specifically, the mirror near your head represents Earth and the light it reflects into space—and toward the moon. With our moon, Earth itself reflects sunlight that brightens the "nighttime" side of the moon. This effect, called **earthshine**, is best seen at the crescent phases. This is sometimes referred to as "the young (or old) moon in the new moon's arms."

The second moon oddity involves the tilt of the crescent as seen over a six- to eight-month period, notably during winter and summer. What do you notice? As seen by comparing Figures 20-2a and 20-2b, the tilt of the crescent shape relative to the horizon changes from season to season, with the greatest change seen between winter and summer. Ancient skywatchers spoke of the winter crescent moon as the "wet moon," filling up with the rains and snows of the season as if it were an upright bowl. But during spring and summer, the crescent shape slowly tilts southward. To some ancients, as the moon's "horns," or **cusps** (the pointed edges of the crescent), shifted to a "pouring" position, the moon would "lose" its water and thus cause the great summer rains. Then, it was called the "dry moon."

It is interesting to note that many early civilizations that used lunar calendars, such as the Hebrews, Chinese, and Moslems, relied heavily on the visibility of the young waxing crescent to mark the start of the month. Many recent clashes in the Middle East have arisen because of questions about the young crescents and the start of the festival Ramadan, a time of fasting in the ninth month of the Moslem calendar.

What's really going on here has to do with the moon's path in the night sky and Earth's position in its orbit, or more simply, what season we are in. If you were to go outside on successive clear nights and make sketches of the moon on a star chart, you would be able to trace out part of its path and notice it moves slightly eastward night to night, relative to background stars. But during the year, as Earth orbits the sun, this path slowly changes its place relative to the horizon (see Figure 20-3) because Earth is gradually pointing either away from or toward the moon's orbital path. During summer, when the northern hemisphere points in the direction of the sun, the night side of Earth is tipped **away** from the moon's orbit. The opposite occurs during winter, when the northern hemisphere points away from the sun and night observers are tipped **toward** the moon's orbit. This causes the moon to appear higher in the night sky during winter and lower in the night sky during summer (see Figure 20-4).

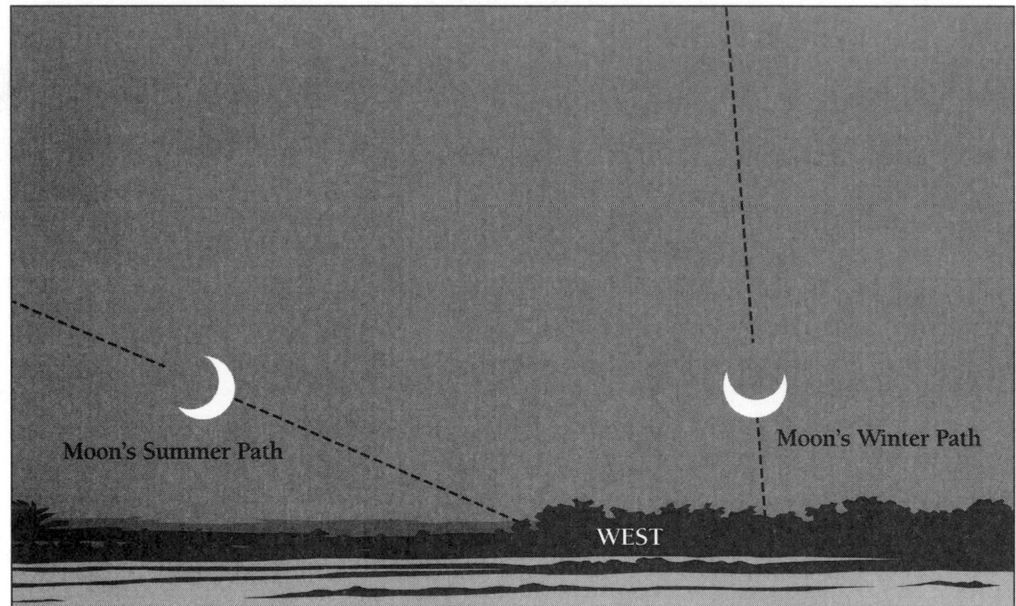

FIGURE 20-3. The paths of the setting crescent moon during summer and winter. Notice that the winter moon sets far north of west, whereas the summer moon sets far south of west.

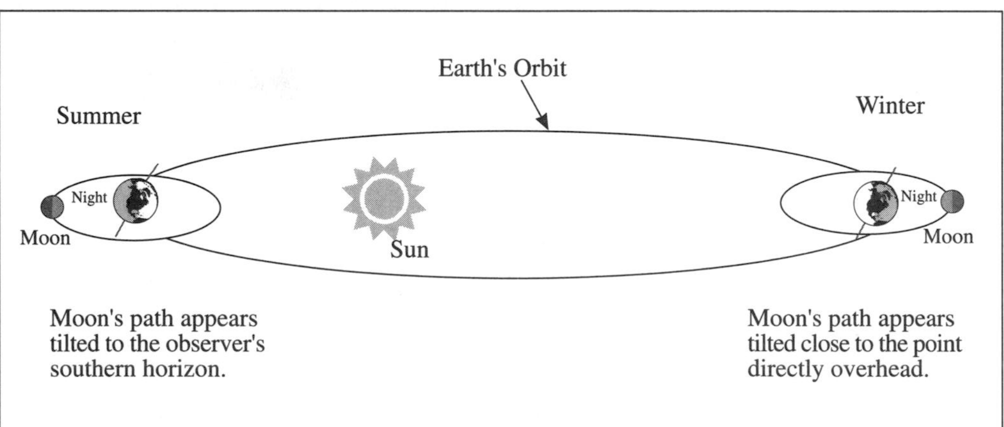

FIGURE 20-4. The path of the moon during night hours changes as a result of Earth's orbit about the sun. During summer, the observer is tipped away from the moon's path, making the moon's path appear closer to the southern horizon. In winter, the night observer is tipped in the direction of the moon, causing the moon's path to be higher in the sky.

In a more familiar way, the sun's annual path in the sky varies greatly as well. During the northern hemisphere's summer, Earth is tilted in the direction of the sun, yielding a longer and higher path of the sun across the sky. As a result, sunlight falls on a more concentrated and smaller portion of Earth's surface, making it hotter. The opposite happens in winter, with the sun's path being lower and shorter as the northern hemisphere tilts away from the sun. With the moon's nightly path, the reverse occurs. During summer evenings it is greatly slanted toward the southern horizon, whereas its winter path stands nearly vertical to the western horizon. Thus during summer you see the long, high path of the sun during the day and the low, tilted path of the moon at night. During winter, the opposite occurs; the daily path of the sun is low toward the south, and the moon's path is high and long at night.

By making sketches of both the moon and sun relative to the horizon over the course of the year you can visualize this constant "dance" that the moon and sun present to Earth-bound observers. Careful observation of the seasonal star charts in Activities 4 through 7 will show that the sun's path, or ecliptic, is low toward the southern horizon during summer nights, but is near the top of the sky during winter nights.

Can you spot this subtle change in the moon as it phases? Can you see lunar craters or seas on the Earthlit side of the moon? Does the Earthshine brightness vary with season or weather changes? On what crescent date do you see it tipped farthest (and least) toward the south? Try making sketches of the changing crescents, month by month, both with the naked eye and through a telescope. You may also want to try snapping a few photos of the crescents or Earthshine with your camera. Consult Activity 48 for tips on shooting astronomical subjects.

21 Observing Lunar Eclipses

Level: All

Objectives: To learn about and observe eclipses of the moon

Materials:

- A night when the moon is in eclipse (see Table 21-1)
- A clear, dark location from which to view the sky
- Binoculars or a small telescope (optional)

BACKGROUND

Imagine yourself living in the distant past. You, like the ancients, might believe that natural events like earthquakes, thunder, and volcanic eruptions were mysterious punishments for wrongdoing. Particularly threatening were eclipses of the sun and moon. The ancient Chinese believed the sun or moon was being devoured by a dragon during an eclipse or that the gods of light and dark were fighting. In 1504 Jamaican natives believed Christopher Columbus held magical powers over the sky when a lunar eclipse occurred after Columbus claimed he could turn the moon a blood-red color in demand for their cooperation.

Eclipses are nothing more than the passage of one celestial body through the shadow of another. When the moon lies between Earth and the sun, a solar eclipse can occur, putting a narrow region of Earth within the moon's shadow (see Activity 27). A lunar eclipse occurs when the moon passes through Earth's shadow, with Earth between the sun and moon. In each case, one object is blocked by another. Both eclipses occur when the sun, Earth, and moon are lined up. During a solar eclipse, the most exciting time lasts but a few minutes, while the dramatic climax of a lunar eclipse can last well over an hour.

Several different kinds of lunar eclipses can occur, depending on the nature of Earth's shadow and the moon's orbit. Earth casts a dark central shadow, called the **umbra**, which trails some 870,000 miles into space and is about six times the diameter of the moon at the moon's distance. Earth also casts a lighter outer shadow, the **penumbra**, which goes even farther into space and is some 10,000 miles across where it intersects the moon's orbit. Two shadows occur because the sun is not a point of light but is instead a large disc with light rays emanating from its top, middle, and bottom. This gives objects two shadows, as you can see by observing the shadows of your hand or of a stick on a sunny day (see Figure 21-1).

FIGURE 21-1. The shadow of a small tube with a lamp as the light source. Notice that the shadow becomes fuzzier farther from the object. Because the lamp, like the sun, is not a point source of light, the shadow has light and dark parts, called the penumbra and umbra.

The moon is always full at the time of a lunar eclipse because Earth will lie between sun and moon—the required configuration for both a lunar eclipse and a full moon. But a lunar eclipse does not occur at each full moon (see Appendix 3). Why not? Imagine Earth as a marble orbiting a basketball (the sun) on a level tabletop. The moon's path does not lie flat on that tabletop but is tilted about 5°. As a result, the lunar orbit intersects Earth's orbit at only two points, called **nodes**, each month. On average, though, the moon reaches a node at the new or full phase only about once every eight months, at which time a lunar eclipse can occur (Figure 21-2).

A **penumbral lunar eclipse** occurs when the full moon passes only through Earth's penumbral shadow, barely darkening the moon (Figure 21-3). These eclipses can last several hours but rarely attract attention from people because the moon barely diminishes in brightness. Sometimes a faint dirty gray may be seen, but don't expect anything spectacular.

More impressive is the passage of the full moon through a part of Earth's umbra, giving a **partial lunar eclipse** (Figure 21-3). Begin by looking for a black, deep brown, gray, or even reddish color on the eastern-facing edge of the moon. Depending on how much of the moon is in the umbra, a partial lunar eclipse can last well over two hours.

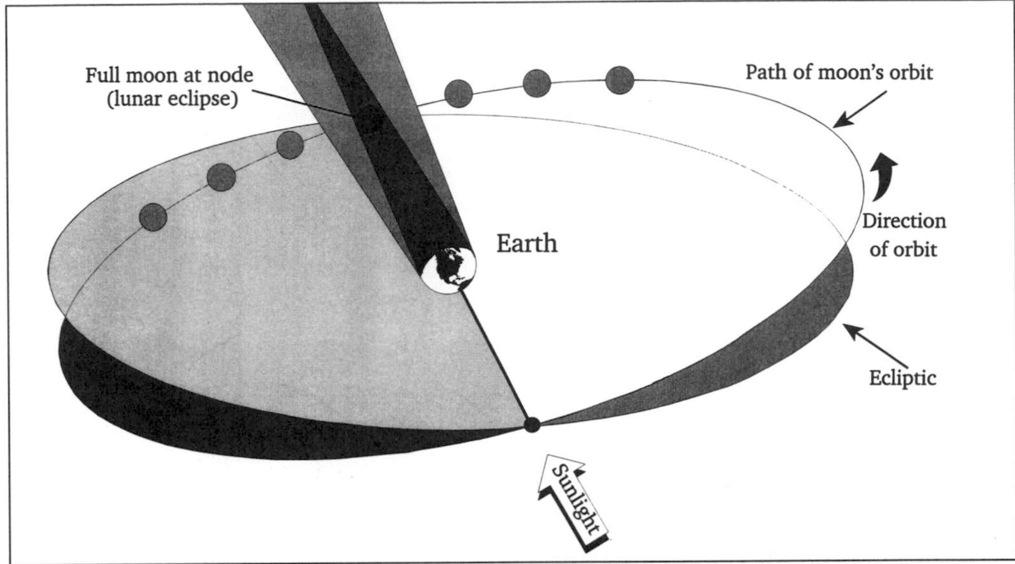

FIGURE 21-2. The paths of the sun and moon shown together demonstrate why eclipses occur only when the sun and moon meet at points common to both paths (called nodes). Note that this diagram shows the placements of the sun and moon from the perspective of an Earth-based observer.

FIGURE 21-3a. A sequence photo showing the moon passing partially through the dark umbral shadow (June 14–15, 1992).

The most dramatic lunar eclipse is the **total lunar eclipse**, in which the full moon passes through the penumbra, then the umbra, and then the penumbra again (Figure 21-4). Duration of the umbral eclipse depends on both the moon's distance from Earth and from the umbra's center. As the moon silently and ominously fades into nothingness, it is easy to imagine our ancestors scurrying about in fright at this awe-inspiring sight.

ACTIVITY

Partial and total eclipses begin with first contact, when the moon just enters the umbra. Watch the moon's eastern-facing (left) edge begin to darken first. In a total eclipse, second contact marks the complete immersion of the moon in the umbra; this is the beginning of **totality**, lasting up to 100 minutes, when the moon may display a variety of unusual colors. Third contact begins the moon's exit from the umbra as the eastern edge of the moon brightens. Totality ends with fourth contact as the entire moon leaves the umbra and completes its second passage through the penumbra (see sequences in "View from Earth" in Figure 21-5). Lasting up to three hours and forty minutes, a total lunar eclipse is visible everywhere on Earth where the moon is above the horizon.

FIGURE 21-3b. Position of the moon during a partial lunar eclipse, in which only part of the moon passes through the dark umbral shadow.

Penumbral Shadow

Moon's orbit

Umbral Shadow

Moon during a partial eclipse

Earth

View from Earth

Penumbra

Umbra

Sunlight

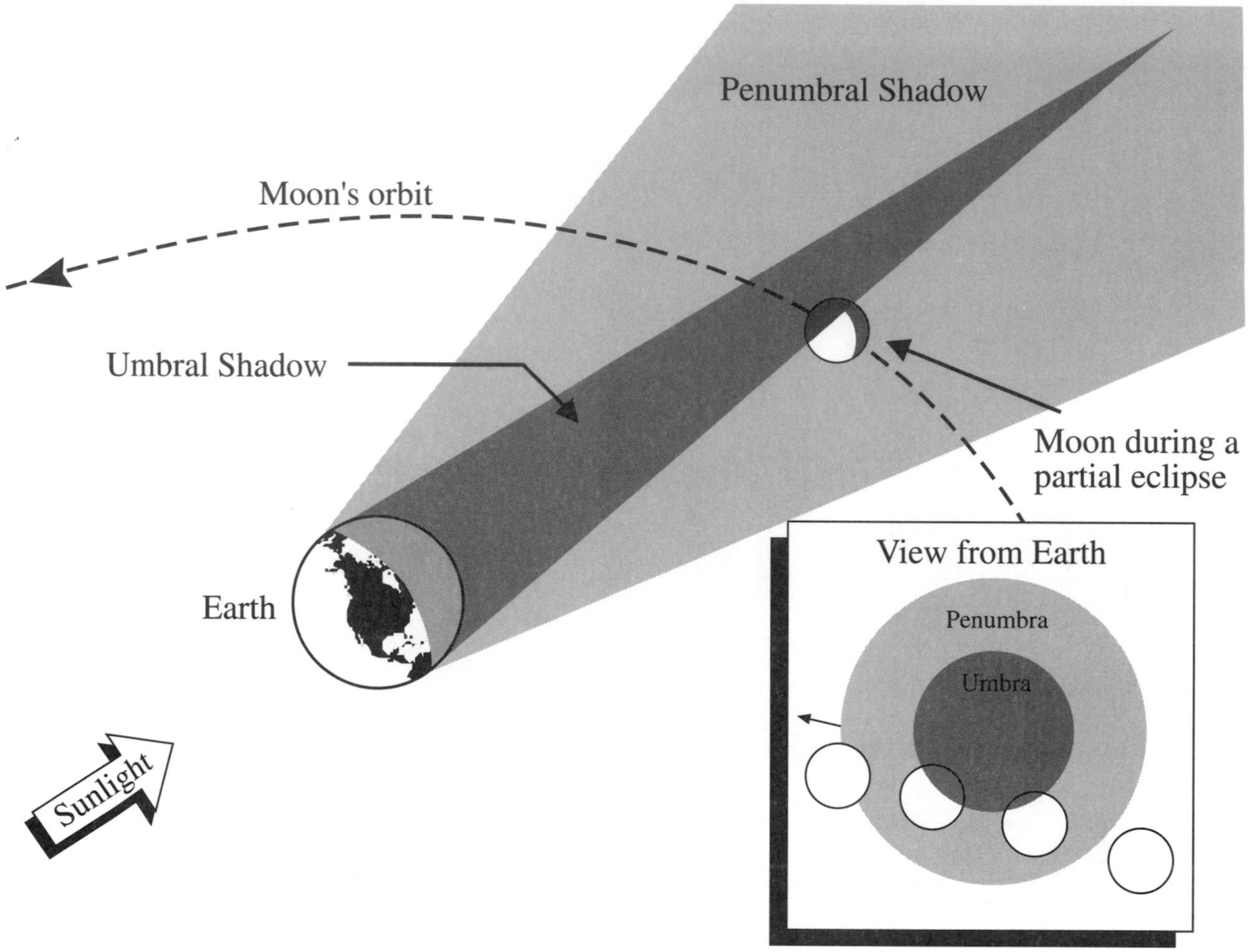

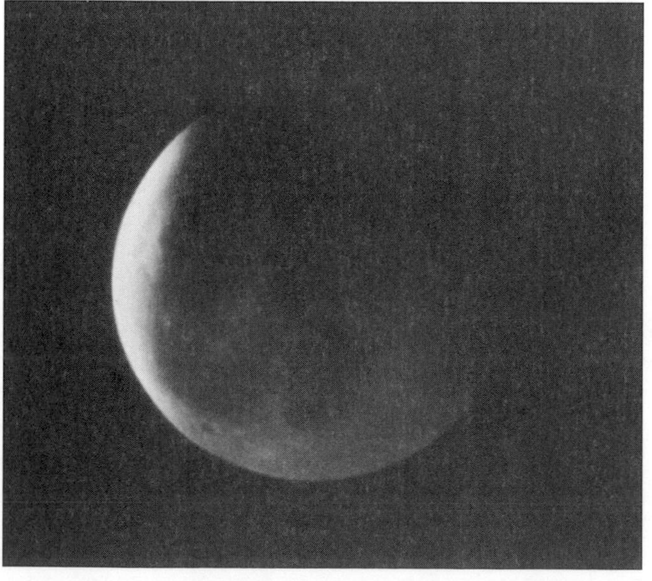

As it enters Earth's umbra, the moon may range in hue from a bright reddish-copper to a deep blue-gray or even black. Earth's atmosphere scatters blue sunlight (giving us a blue sky) and bends, or **refracts**, red sunlight. Indirect sunlight filtered through and refracted by our atmosphere can cast an orange-red glow upon the moon as it passes through the umbra, most prominently when the atmosphere is cloud-free and dust-free. With the addition of clouds, rain, or snow and small particles from pollution, meteors, and forest fires, more sunlight can be blocked,

FIGURE 21-4. A photo showing the moon as it passes completely through the dark umbral shadow, yielding a total lunar eclipse. Notice how dark the moon is (December 9, 1992).

FIGURE 21-5. Position of the moon during a total lunar eclipse, in which the moon passes entirely through the dark umbral shadow.

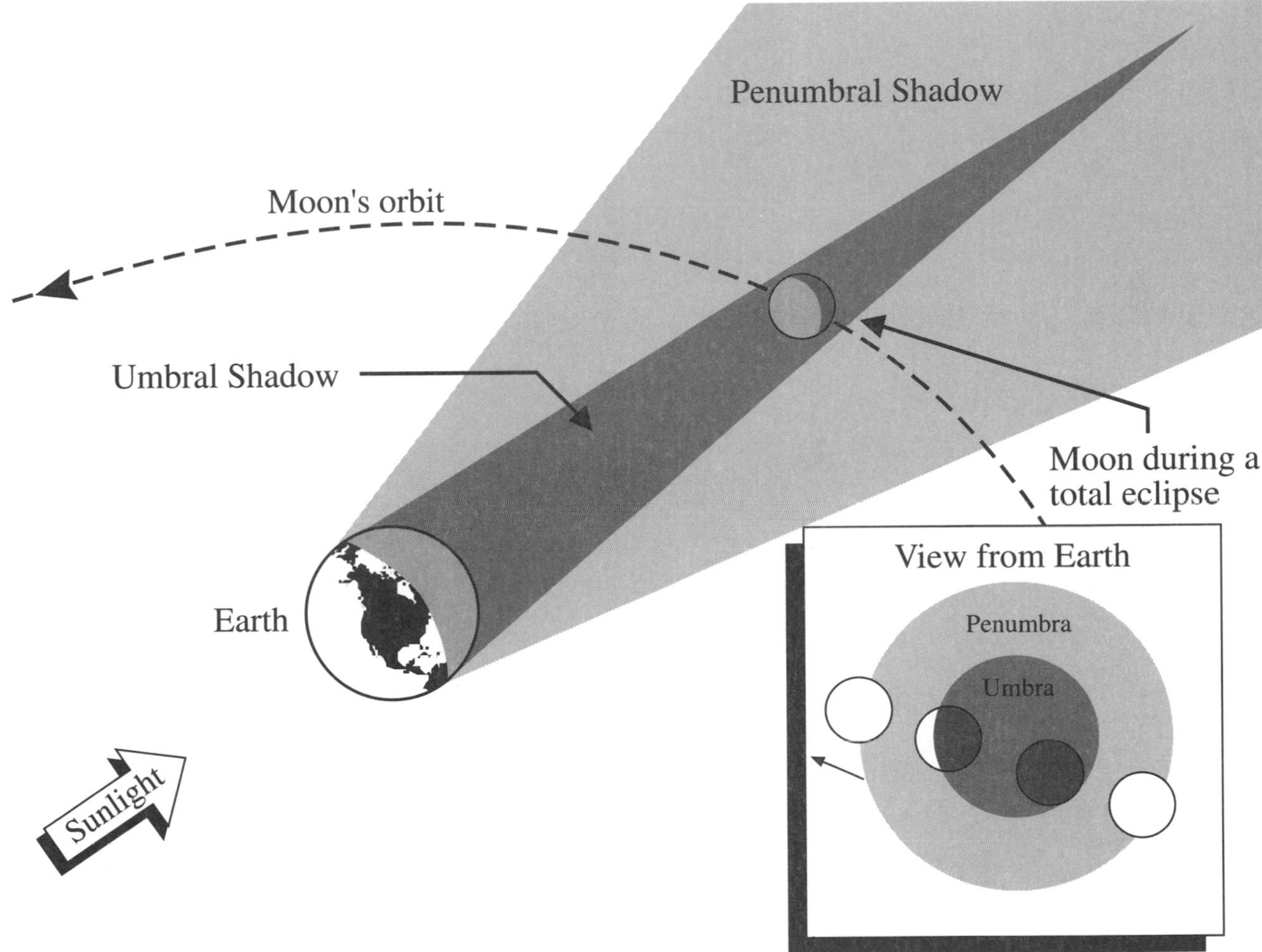

darkening the moon. Volcanic eruptions can spew tons of ash and finer particles into the air, darkening eclipses for years. The eclipses of December 30, 1963, and December 30, 1982, were called "black" eclipses due to the eruptions of Bali's Mount Agung and Mexico's El Chicon, respectively. The partial and total lunar eclipses of 1992 saw the moon virtually disappear, then reappear, a result of the June 1991 eruption of Mount Pinatubo in the Philippines.

Watching a lunar eclipse is safe and, unlike looking incorrectly at the sun, poses no threat to your vision. A lightweight pair of low-power binoculars (7 x 35 or 7 x 50) will enhance your view considerably; larger binoculars will be harder to hold steady. Try rating the darkness of the eclipse (from complete black to a bright blood-red) at the very middle of the eclipse, indicating actual color variations of Earth's shadow. Can you see craters or maria during totality? If you are observing the event far from city lights, you may be surprised at the increased number of stars visible during totality. Try to estimate the faintest object you can see before, during, and after totality. To plan your next eclipse, check Table 21-1 for upcoming events. Happy eclipse watching!

TABLE 21-1

The following is a list of penumbral, partial, and total lunar eclipses, 1994–2001.

DATE	TYPE OF LUNAR ECLIPSE
May 24, 1994	Partial
November 18, 1994	Penumbral
April 15, 1995	Partial
October 8, 1995	Penumbral
April 3–4, 1996	Total
September 27, 1996	Total
March 24, 1997	Partial
September 16, 1997	Total
March 12–13, 1998	Penumbral
August 7–8, 1998	Penumbral
September 6, 1998	Penumbral
January 31, 1999	Penumbral
July 28, 1999	Partial
January 21, 2000	Total
July 16, 2000	Total
January 9, 2001	Total
January 30, 2001	Penumbral
July 5, 2001	Partial

SECTION III

The Sun

22 How *Not* to Look at the Sun

Level: All

Objective: To learn how to study the sun safely

Materials:

- A telescope or binoculars mounted on a tripod or other support
- A piece of white paper or cardboard

BACKGROUND

What is the sun? To most people, it is a strange, extremely bright "thing" in the daytime sky that gives off heat and light. Other people know that staying out under it for long periods of time will cause damage to our skin and that staying under it for too long will result in severe burns. To most, that is where their knowledge of the sun ends.

Very few people think of the sun as a star, but that is exactly what it is. In fact, the sun makes a great "model star" because it is typical in every regard to many of the thousands of stars seen only at night.

How can this be? If it is like the stars at night, then why does it appear so much bigger and brighter? To answer this question, we must think in terms of distance. The stars at night are so far away that their distances cannot be measured conveniently in miles. Instead, astronomers use the **light-year**, where one light-year equals the distance a beam of light travels in one Earth year. Racing through the vacuum of space, a light beam covers 186,000 miles in only one second—or almost six trillion (6,000,000,000,000) miles in one year. Yet even traveling at that unimaginable speed, starlight takes years (in some cases, **tens, hundreds**, or **thousands** of years) to reach us here on Earth because of the stars' incredible distances away.

Our star, the sun, is much closer. It is so close that its distance cannot be measured in light-years, but instead in light-**minutes**. The sun is about eight-and-one-third light-minutes away, or about 93 million (93,000,000) miles. This means that the sunlight reaching Earth right now left there just over eight minutes ago.

Because of the sun's proximity, astronomers use it to learn more about the stars—and so can we. But be forewarned that looking at the sun can be dangerous! The sun's ultraviolet rays, the same rays that cause sunburn, will burn your eyes' sensitive retinas much faster than they will your skin. Without proper precautions, permanent eye damage, even blindness, will result. That is why we must pause now to discuss how—and how NOT—to look at the sun.

ACTIVITY

First, the DON'Ts. Most importantly, DON'T ever look at the sun directly with your eyes, not even for a second! Also, DON'T ever use sunglasses, welder's glass, smoked glass, or overexposed film. They can all lead tragically to blindness. Some telescopes come with sun filters that screw onto the eyepiece, but DON'T ever use them, as these are the most hazardous of all. Although the filter will reduce the sunlight to a safe level, these glass filters have been known to crack under the focused heat of the sun. When that happens, unfiltered sunlight will burst through the telescope and into the unwary observer's eye.

All this is not meant to scare you away from viewing our star, but only to make you aware of the need for proper precautions. The safest (and cheapest) way to view the sun safely is to project its image through a telescope or binoculars onto a piece of white paper or cardboard, as shown in Figure 22-1.

First mount the telescope or binoculars on a tripod or other rigid support. (If a tripod is not available, try to secure the telescope or binoculars in the V of a tree branch). Remember, DO NOT look through the telescope or binoculars at any time, not even to aim them. You can align the instrument by adjusting its shadow until it is shortest. (If using binoculars, keep a dust cap over one of the front lenses to prevent two overlapping images; if using a telescope, keep the finderscope capped to prevent someone from looking through it accidentally.)

Once centered on the sun, turn the focusing knob in and out until the image on the cardboard is sharp and clear. The size of the sun's image can be reduced or enlarged by moving the cardboard in and out (the instrument will have to be refocused each time the distance changes).

Now that you know how to look at the sun safely, try the next activity. There you will discover how to make valuable scientific observations in order to learn even more about our star.

FIGURE 22-1. Author Harrington's daughter Helen demonstrating how to view the sun safely.

23 Estimating the Sun's Period of Rotation

Level: All

Objective: To learn how long it takes the sun to rotate on its axis

Materials:

- A telescope or binoculars mounted on a tripod or other support
- Photocopies of Sun Observation Form (included with this activity)
- A piece of cardboard
- Masking tape
- Pencil

WARNING: Do NOT perform this activity before you understand how to view the sun safely. Viewing the sun without adequate eye protection can result in permanent blindness. Review Activity 22 to learn a method of viewing the sun safely.

BACKGROUND

For the past 4.5 billion years, the sun has been bathing Earth with life-sustaining heat and light. Yet for all its majesty, the sun is just an average star. Measuring 864,000 miles in diameter with a surface temperature of 11,000° F, our star is classified by astronomers as "type G"—a yellow **dwarf**!

The sun is composed of three basic layers. The brightest is the **photosphere**, the visible surface of the sun (Figure 23-1). Through large, specially equipped telescopes, the photosphere has a pebbly look. This results from rapidly rising pockets of gaseous material. As it is heated deep within the solar core, at a temperature of 270 million degrees Fahrenheit, gas is sent rushing to the top of the photosphere. This appearance, like what you see when bubbles rise to the top of a pot of boiling water, is called **granulation**. Each granule, representing a bubble of heated gas, may measure from about 200 to 1,000 miles across—larger than many states.

Far easier to view are small, dark patches scattered across the surface of the photosphere, called **sunspots**. Sunspots are transient features that apparently result from powerful magnetic fields that reduce some of the outwardly pouring radiation from the sun's core. The temperature of the affected region is about 3,000° F cooler than the surrounding photosphere, causing the sunspot to appear darker by contrast.

Each sunspot consists of a black central portion, the **umbra**, and an encircling grayish area called the **penumbra**. Sunspots range in size from one granule to thousands of miles in diameter.

The number of sunspots visible can vary greatly over a period of time. At times there might be a hundred or so spots observed; at others, the solar disk may be nearly devoid of them. By the middle of the nineteenth century, astronomers had plotted a mysterious eleven-year cycle to sunspot activity. Every eleven years the number of sunspots reaches a maximum, with an ebb midway between. Nobody really understands why.

Sunspots reached a maximum in 1959, 1970, 1981, and 1991. The next maximum is expected in late 2001 or early 2002. As we approach the next solar maximum, the number of sunspots visible is expected to increase with each passing year. As sunspot numbers continue to increase, so will the frequency of SOLAR FLARES. Solar flares are mammoth eruptions from sunspots that emit charged particles from deep within the sun's interior. These charged particles race outward into the solar system, colliding with anything that gets in their way. As the particles bombard the Earth, they are drawn toward our planet's magnetic poles, causing the aurora, or northern lights. See Activity 24 for more details on this fascinating sight.

Above the photosphere is the **chromosphere**, a relatively thin solar layer consisting of millions of spike-like projections called **spicules**. Normally, the chromosphere is invisible due to the overwhelming brightness of the photosphere. It was not discovered until 1842, when astronomers first spotted its deep red ring encircling the blocked photosphere during a total solar eclipse. Today, astronomers monitor the chromosphere daily using a special device called a coronagraph.

Still higher above the sun's visible disk is the solar atmosphere, or **corona**. The corona is a ghostly white halo of intensely hot hydrogen that engulfs the sun and may be observed only during a total solar eclipse (see Activity 27).

ACTIVITY

You probably already know that the Earth turns on an axis that passes from the north pole to the south pole. It takes twenty-four hours for our planet to rotate once on its axis. That's what gives us day and night. While the sun spins on an axis, too, it takes much longer than Earth to turn completely around.

The seventeenth-century astronomer Galileo was the first to discover that the sun rotates when he turned his first crude telescope toward the sun in 1610. From his observations of sunspot groups and their motion, Galileo concluded that the sun rotates once about every four weeks, a figure that is still generally accepted today. In this activity, you will confirm for yourself Galileo's discovery.

Begin by making a photocopy of the Sun Observation Form and taping it onto a piece of stiff cardboard. Shine the sun through your telescope or binoculars (see Activity 22), adjusting the distance from eyepiece to screen until the sun's disk fills the circle. Watch as the sun slowly moves from east to west across the screen and mark the direction of movement. (Remember, the sun isn't actually moving in our sky; this apparent motion is due to Earth's rotation.) Rotate the Sun Observation Form until the east-west motion of the sun in the sky matches the east-west axis drawn on the form.

With the actual sun and the observation form aligned with each other, draw the exact positions and sizes of all the sunspots visible. Be as precise as possible. (Attaching the form to a fence or side of a house will make it far easier to sketch.)

On the next sunny day, go back outside with a new copy of the form and again carefully draw the locations and sizes of all visible sunspots. Once back inside, compare the two drawings. What conclusions can be made? It should be apparent that the sunspot patterns are no longer in the same places; they have moved. How much have they moved? In what direction did they move?

Repeat the procedure again on the next sunny day. Then, returning indoors, compare all three drawings. Can you see a pattern of sunspot movement developing? The sunspot groups appear to be moving parallel to the sun's equator because of the sun's rotation.

Draw the sunspots on each sunny day for a two-week period. If you are able to track a sunspot across the sun's disk from one edge to the other, you will find that it takes just about that long for the spot to complete the trip. Since only half of the sun can be seen at any one time, you will have recorded the sun through one-half of a full rotation. So Galileo was right: The sun rotates once about every four weeks. Actually, since the sun is not a solid body but gaseous in nature, different latitudes rotate at different rates. The sun's equator takes about twenty-five days to complete one rotation, as compared with about thirty-four days for the slower-moving poles. Can you estimate where the sun's equator and poles are located in your drawings?

Sun Observation

Date: ______________________ Time: ______________

Telescope: ______________________ Power: ______________

Number of sunspots: ______________________________

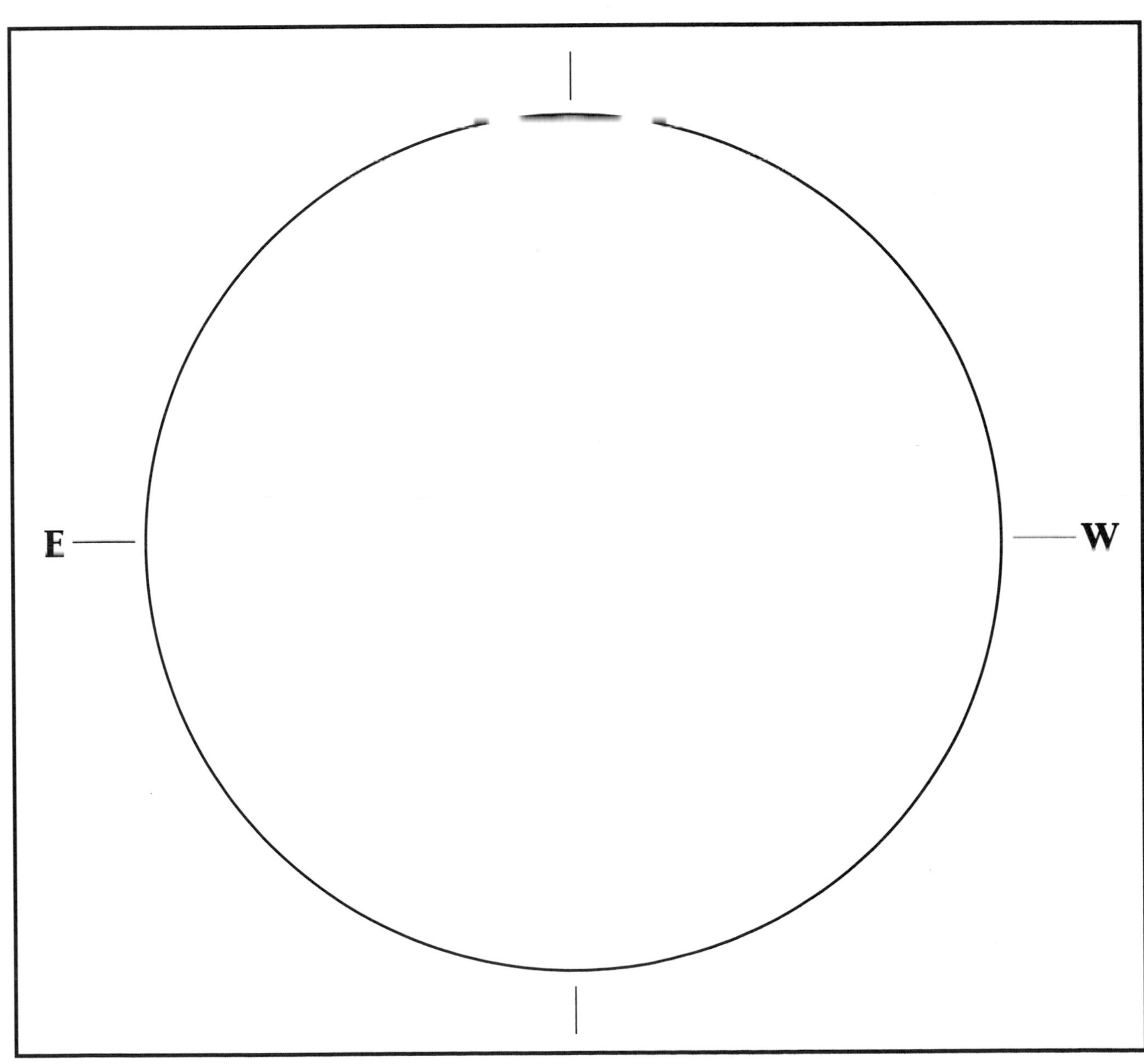

24 Electricity in the Sky

The Northern Lights

Level: All

Objective: To observe and understand the appearance of the Northern Lights

Materials:

- A clear, dark sky
- A little luck (OK, a lot of luck!)

BACKGROUND

Have you ever been listening to your favorite FM radio station when suddenly it fades away and another station, hundreds of miles away, comes through instead? This effect, lasting from a few hours to a couple of days, will finally subside and your local station will return.

It is not unusual to experience such radio and television communications interference during a period of strong sunspot activity. Recall from Activity 23 that sunspots come and sunspots go, but that greater numbers appear during the peak of the as-yet-unexplained eleven-year sunspot cycle. As the number and size of sunspots increase, the conditions that produce **solar flares** also improve.

Solar flares are mammoth eruptions from sunspots that emit charged particles from deep within the sun's interior. When the stream of charged particles flowing outward from the sun, called the **solar wind**, reaches Earth, the particles are either diverted or trapped by Earth's magnetic field. The trapped particles spiral along the magnetic field's lines of force until the particles interact with atoms and molecules in our upper atmosphere, disrupting many forms of communications.

These particles also produce the beautiful **aurora**—the **Northern Lights** in the northern hemisphere and the **Southern Lights** in the southern hemisphere. Auroral displays occur most frequently near Earth's magnetic poles because that is where the field lines enter our planet's upper atmosphere, though they have been seen much closer to the equator, in such locations as Miami and Honolulu.

There is no way to predict exactly when the next aurora will be seen. In general, they occur whenever sunspot activity is high. Whenever you notice increased sunspot activity or unusual radio or television reception, be sure to check the sky that evening and the following few evenings for an aurora. A display may start and remain as merely a dim glow seen in the north, or it may suddenly erupt into a spectacular show of softly colored wavering patterns of light. Regardless of their intensity, auroras are best viewed with just the unaided eye.

Your odds of seeing an aurora depend greatly on where you are on Earth in relation to the magnetic poles. The Northern Lights are most common in an oval region spanning parts of Canada, Alaska, Greenland, Siberia, and the Arctic. In this zone, an aurora is seen about fifteen to thirty nights a year. By contrast, an aurora may be seen only one or two nights a year from the southern United States. But even the brightest aurora will be obscured if the surrounding sky is brightened by lights from nearby cities and towns. That's why it is important to view the aurora from the countryside.

There are five types of auroras. By far the most common is the **auroral patch**. An auroral patch appears as a small grayish glow on or near the northern horizon. Many people misinterpret an auroral patch as a patch of clouds or the glow of a distant city .

How can you tell when you are seeing a bona fide auroral patch and not just a passing cloud or the glow from a distant city? One way to distinguish an aurora from a cloud is to watch for stars through the glow. Stars will shine through an aurora but not usually through a cloud. If you suspect that the eerie glow you are seeing is an aurora, watch it for at least five minutes. An auroral patch usually will not drift east or west, though it may appear to grow or diminish in intensity. Clouds, on the other hand, will float with the prevailing winds but will probably not change dramatically in shape. Also take a look for any clouds silhouetted in front of the glow. If the clouds appear dark in front of the glow, then it is probably an aurora; otherwise the glow has a more earthly origin.

Telling the glow of a city apart from the glow of an aurora is usually just a matter of knowing where the nearest city is. But if you are unfamiliar with the direction to cities in your area, just remember that the glow from an aurora will slowly change in appearance, while the glow from city lights will not.

FIGURE 24-1. Ray aurora. (Photo by Susan and Alan French)

Figure 24-2. Auroral arc. (Photo by Susan and Alan French)

FIGURE 24-3. Veil aurora. (Photo by Susan and Alan French)

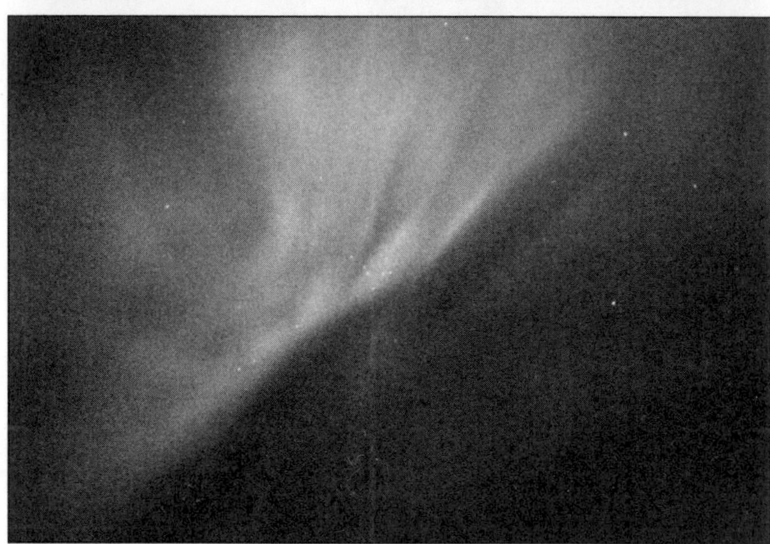

FIGURE 24-4. Auroral corona. (Photo by Susan and Alan French)

Other types of aurora are a little more obvious. A **ray aurora** (Figure 24-1), formed along the lines of force in Earth's magnetic field, appears as vertical shafts of light shining above the horizon like perfectly straight searchlights. An **auroral arc** (Figure 24-2) displays gently curving strips of light that arch across the sky like ghostly ribbons (and are not to be confused with aircraft contrails). An **auroral band** looks similar to an arc except that it is highlighted by intricate kinks and folds. A **veil aurora** (Figure 24-3) looks like an eerie curtain covering a large portion of the sky. Intense aurora displays are highlighted by an **auroral corona** (Figure 24-4) near the zenith. A corona (meaning "crown") appears as a series of converging rays high overhead. In reality we are looking straight up along magnetic field lines.

Brighter displays will show intense colors. Exactly which colors will be seen depends on how high the aurora is in Earth's atmosphere. Most auroras fluoresce between 60 and 150 miles above the surface. At this altitude, the charged solar particles cause oxygen atoms to emit a greenish color. If the bottom of the display is below 60 miles, blue and red colors result from glowing nitrogen atoms.

High-altitude auroras—that is, displays at altitudes between 150 and 600 miles—appear reddish because of oxygen radiating in the red end of the spectrum (rainbow).

ACTIVITY

Keep on the lookout for the next aurora. By monitoring the changing appearance of the sun (as outlined in Activities 22 and 23), you may be able to predict a day or two in advance when the next display will occur. As we progress toward the next sunspot maximum in 2001 or 2002 (see the earlier discussion in Activity 23), the chances of seeing an aurora should increase. Skyline, a telephone message from *Sky & Telescope* magazine, is updated at least weekly with astronomical news, including announcements of solar flares; call (617) 497–4168. Some news broadcasts will mention that a solar flare has occurred. If so, there might be an aurora within two nights.

When you suspect an aurora blossoming, watch it carefully to make sure it actually is an aurora and not just an unusual cloud. Remember, stars can be seen through an aurora but not usually through clouds. An auroral display may last only half an hour or up to three days. If it is an aurora, take careful notes, answering the questions on the Aurora Observation Form included with this activity.

Seeing your first aurora is an exciting event, especially if it is an intense display. If it is especially vivid, the aurora may bathe the entire sky in a wavelike oscillating glow of faint colors. The memory of such an event will last you a lifetime.

Aurora Observation

Date:________________ Your Location:______________________________

What time did you first notice the aurora?____________________________

Do you see any colors? If so, which?____________________________

____

How is the aurora moving?____________________________________

What type of aurora is it?____________________________________

Can you make out any shapes, such as curtains or rays?____________________

____

____

What constellations did the aurora appear in?__________________________

____

What time did the aurora end?________________________________

25 Making a Sundial

Level: Intermediate and advanced

Objective: To learn the basics of telling time using a sundial

Materials:

- A protractor (your own or the one in Activity 12)
- A piece of cardboard, about 6 inches square
- For a vertical sundial, a thin piece of wood, about 8 inches by 12 inches
- For a horizontal sundial, a thin piece of wood, 12 inches square **or** in a 12-inch-diameter circle
- Thick tape, a pencil, ruler, and scissors

BACKGROUND

Since antiquity, skywatchers have observed the heavens with a desire to know, understand, and predict celestial events. Great monolithic monuments such as Stonehenge in England and the Medicine Wheel in Wyoming served as celestial calendars, marking the beginning or end of a specific time of year. These structures served well as annual timepieces, but daily timekeeping, marked by the paths of the sun, moon, or specific stars, was reserved to a more compact instrument. Ancient devices that performed daily timekeeping are called **sundials** (Figure 25-1) and are believed to have originated some 4,000 years ago. In this activity, you will have an opportunity to construct one (or both) of two types of sundials: horizontal and vertical.

Figure 25-1. An example of one of many elaborately decorated wall sundials found today. This one, at New York's Vanderbilt Museum, is some 20 feet high.

Sundials range in size from the ancient "pocket" sundial (somewhat analogous to today's wristwatches) to a sundial in the town of Jaipur, India, that stands 56 feet high. A sundial consists of two main parts. The first part is the **dial face**, inscribed with hours of the day. These usually run from 6 A.M. to 6 P.M. or from 5 A.M. to 7 P.M. The second part is a piece of metal or other strong material placed in a vertical position on the dial face. Upon closer inspection, you'll notice that this vertical piece, called a **style** or **gnomon** ("no-mon"), is slanted. The angle of the slant is the same as the observer's latitude.

For example, if you are using the sundial at latitude 50° north, then the gnomon angle must be 50°. A sundial is positioned so that the shadow cast by the gnomon on a sunny day falls on the face at a point that marks the time of day.

ACTIVITY

Use a protractor to measure and mark your latitude on the piece of cardboard, as shown in Figure 25-2. Cut the cardboard as shown. This piece will serve as the gnomon for the sundial. Next, photocopy the dial face of your choice (Figure 25-3 or 25-4) for either a horizontal or vertical sundial. If you choose to make the horizontal dial, use either the circular or the square piece of wood indicated in the materials list. Cut out the dial face and tape it to the piece of wood. If you choose to make the vertical dial, cut out the vertical face and tape it to the 8-inch by 12-inch piece of wood. Tape the gnomon to the dial face (Figure 25-5) and make sure it stands as vertical (or perpendicular) as you can make it. Now you are ready to make use of the dial as a timepiece.

The sundial must be situated a certain way. For the vertical sundial, place it on a wall or a post so that the front of the dial faces due south. You can usually do this only roughly by first hanging a small weight (such as a key on a string) around the gnomon and setting up

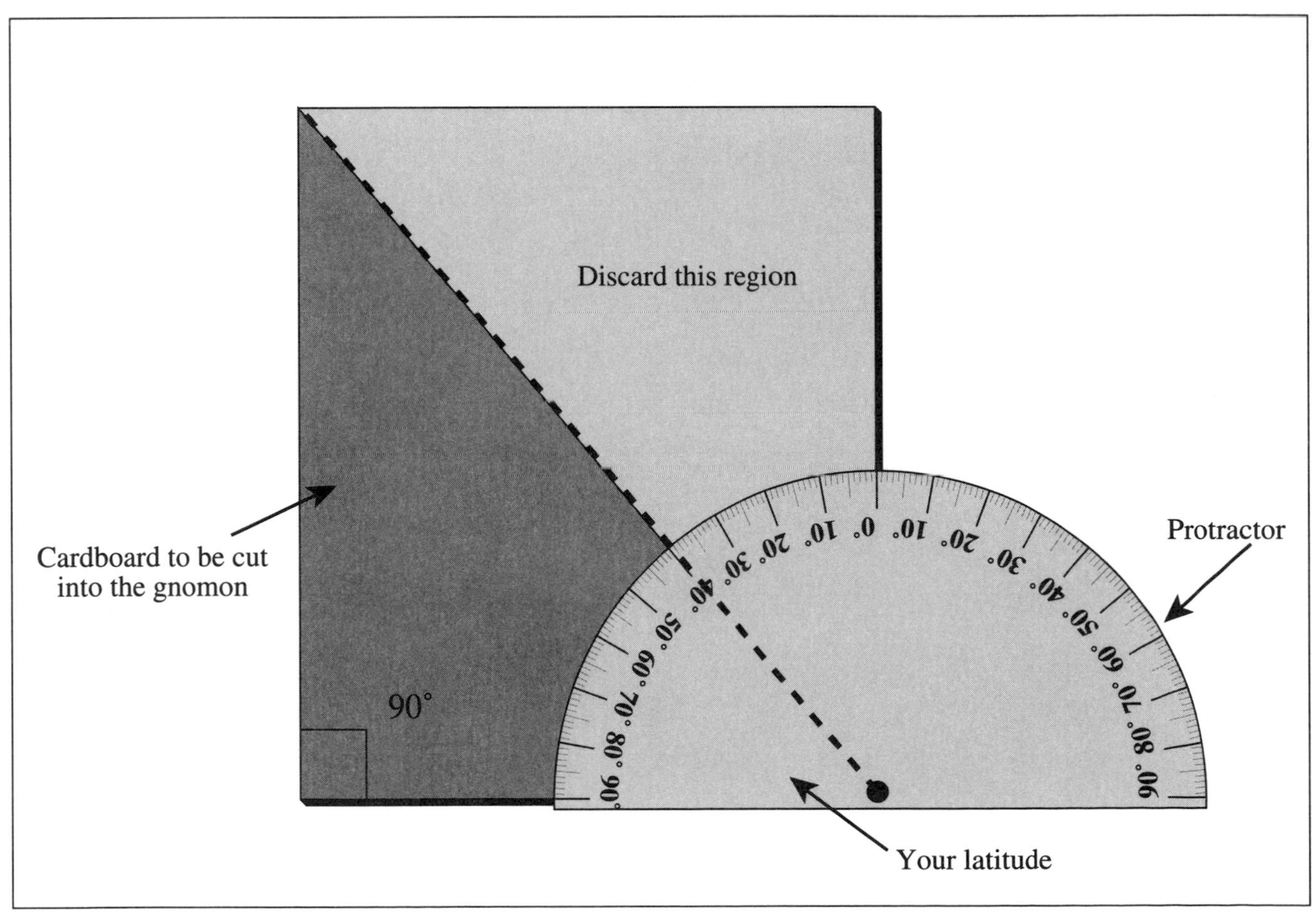

FIGURE 25-2. Using a protractor, pencil, and ruler, mark off the angle of your latitude onto the cardboard. For example, if your latitude is +40°, the gnomon's angle will also be +40°.

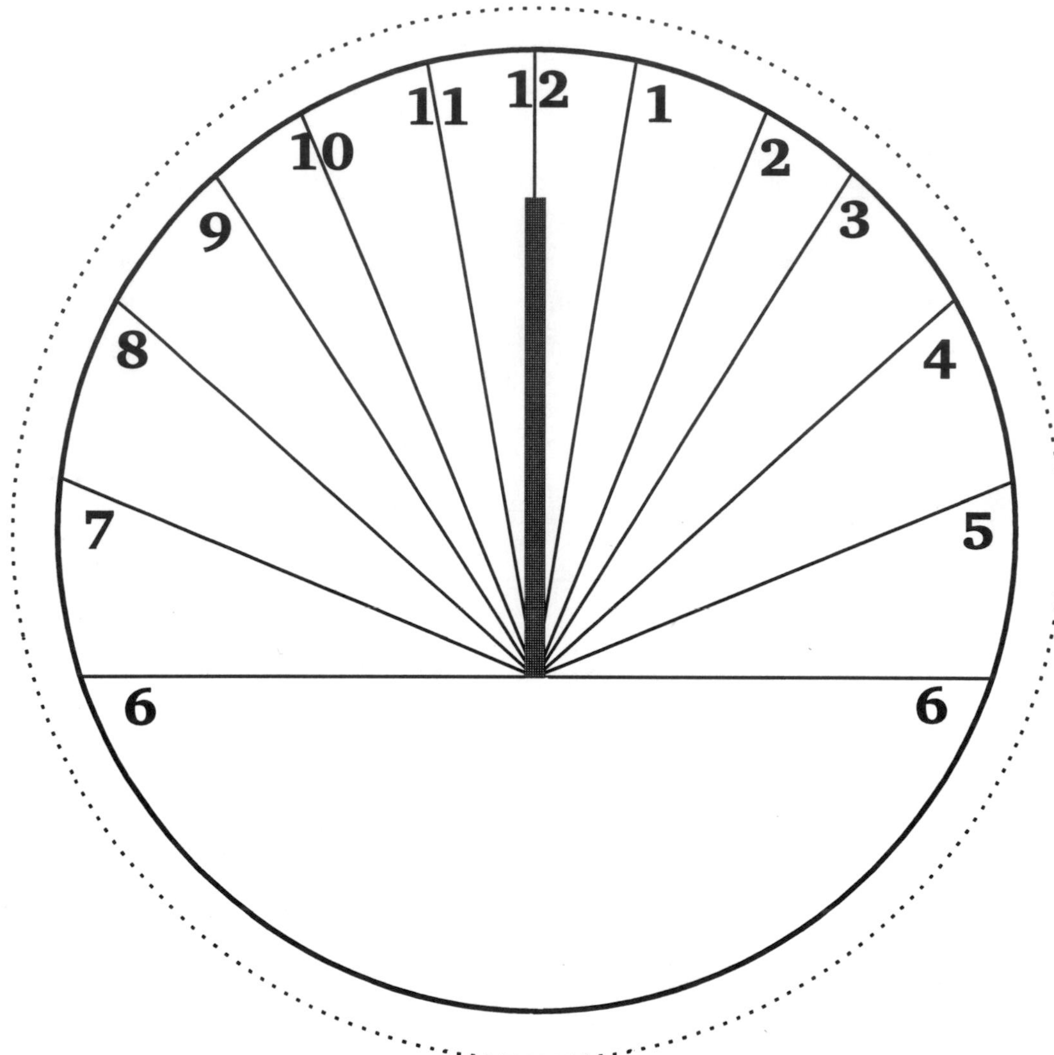

FIGURE 25-3. Horizontal sundial template. Photocopy, cut along the outer dashed line, and tape the gnomon to the vertical strip along the "12" line.

the sundial at exactly 12 noon. Keep the sundial precisely vertical at all times. Face the dial toward the sun and turn it until the gnomon's shadow lines up with 12 noon. Now attach the sundial to the wall or post. Your sundial is now in operation.

For the horizontal sundial, use a compass or the North Star to determine which direction is north (see Activities 2, 3, and 12). Align the gnomon to north as shown in Figure 25-5, with the sundial resting on a flat, level surface. This sundial is now in action.

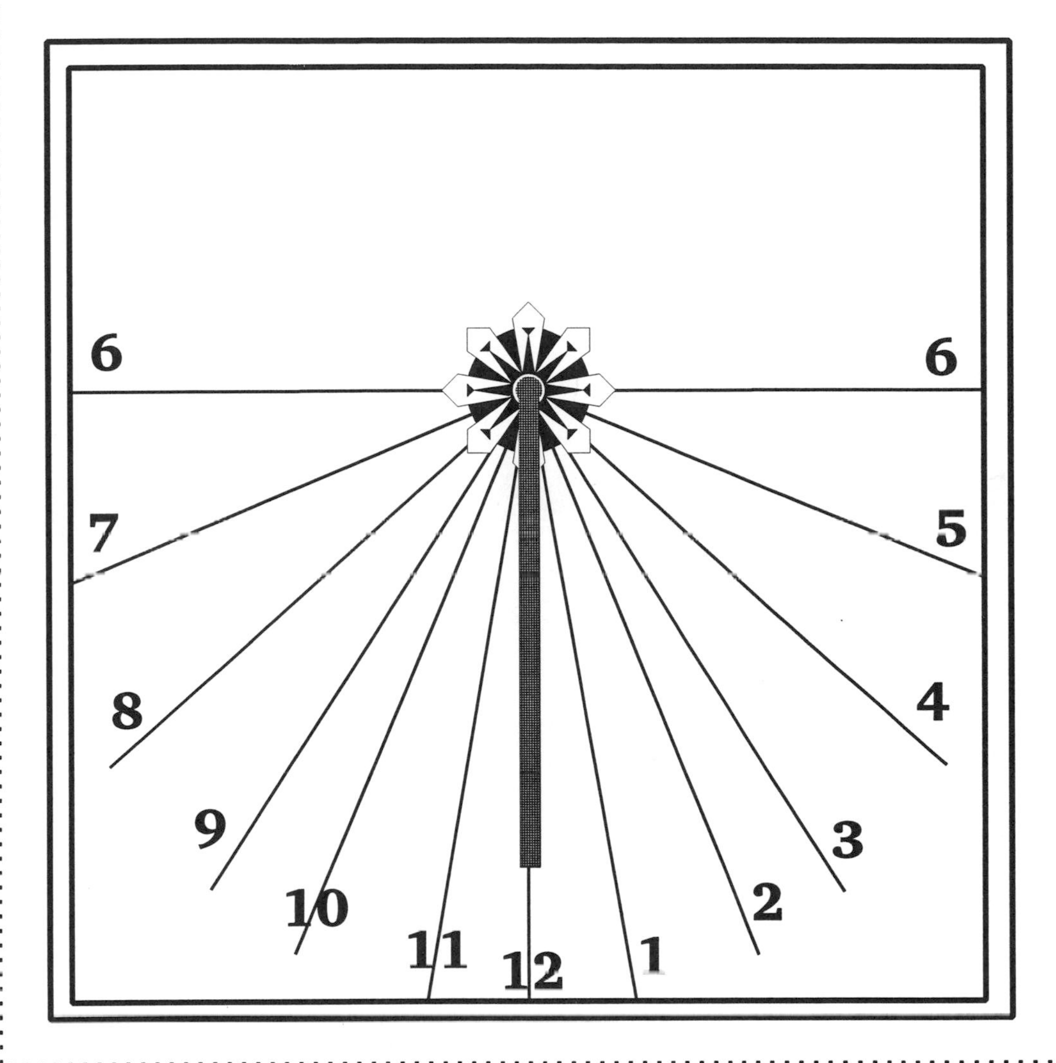

FIGURE 25-4. Vertical sundial template. Photocopy, cut along the outer dashed line, and tape the gnomon to the vertical strip along the "12" line.

Watching the shadows cast by the gnomon during the day can be instructive and can help you understand the importance the ancients placed on daily timekeeping. Throughout the world today, lives are focused around the movements of clocks and watches, the descendants of the simple sundial.

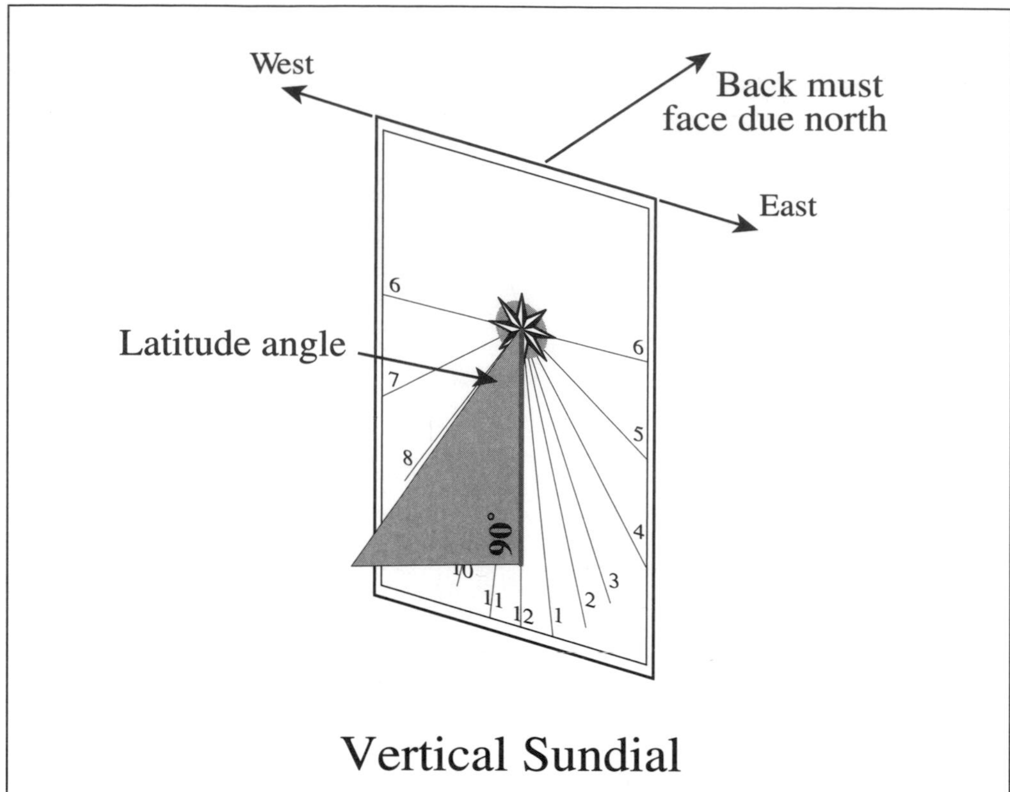

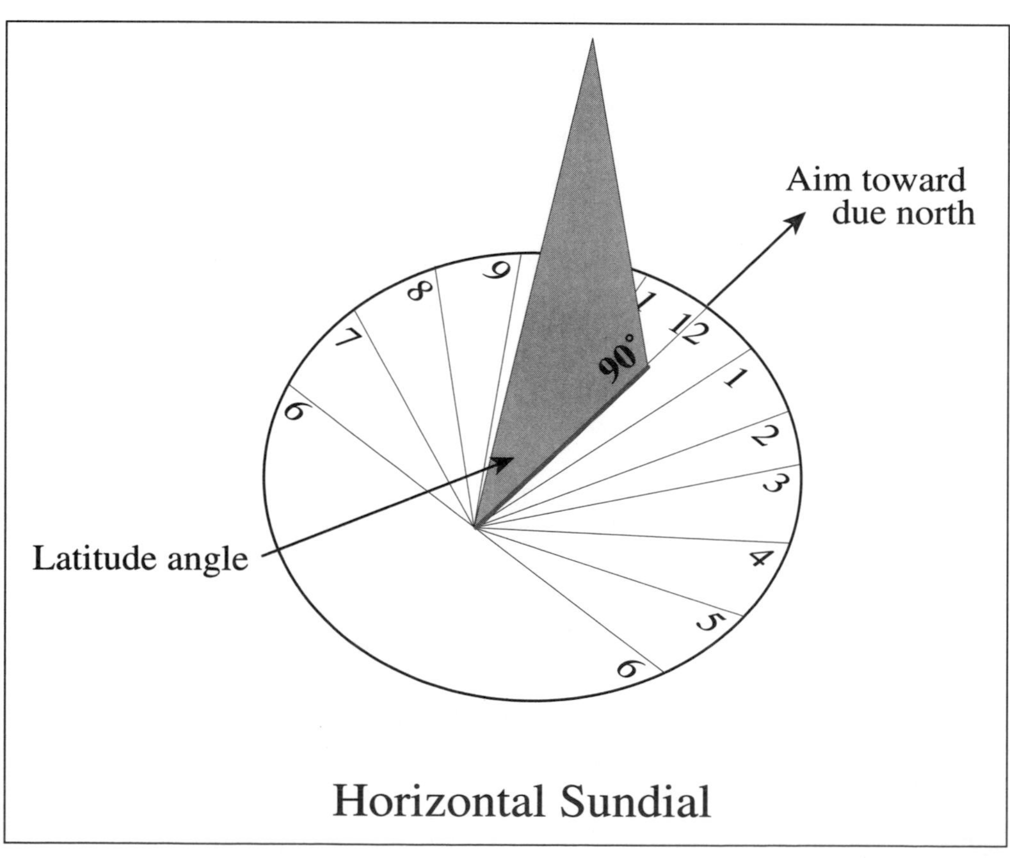

FIGURE 25-5. Aligning the sundials to north. Use a compass to find approximate north or sight on Polaris at night (see Activities 1 and 3).

26 Figure Eight in the Sky

Making an Analemma

Level: Advanced

Objective: To see how the sun, as a "clock," does not always agree with your watch

Materials:

- A piece of half-inch-thick plywood, 16 inches by 30 inches
- Two 3/4-inch dowels, one 20 inches long and one 12 inches long
- A drill and a 3/4-inch bit
- Sandpaper
- Hammer
- A piece of plain white paper, about 2 feet square
- A 32-inch length of string
- Tape

BACKGROUND

Old globes of the world (Figure 26-1) included an elongated figure-eight known as an **analemma**. This figure was precisely placed on the Pacific Ocean, directly between the latitudes +23.5° (the Tropic of Cancer) and –23.5° (the Tropic of Capricorn). Originally used to determine the position of the sun in the sky from any location on Earth for any date, this figure has been omitted from globes because of its apparent complexity. Too bad; it's fascinating! It is a representation of the actual position of the sun in the sky at the same time each day for an entire year. A multiple-exposure photo of the sun taken at the same time of day over the course of a year will reveal this odd-looking shape (Figure 26-2). Why does the analemma have this shape? To uncover this mystery, it will be necessary to understand how Earth orbits the sun and how astronomers keep time.

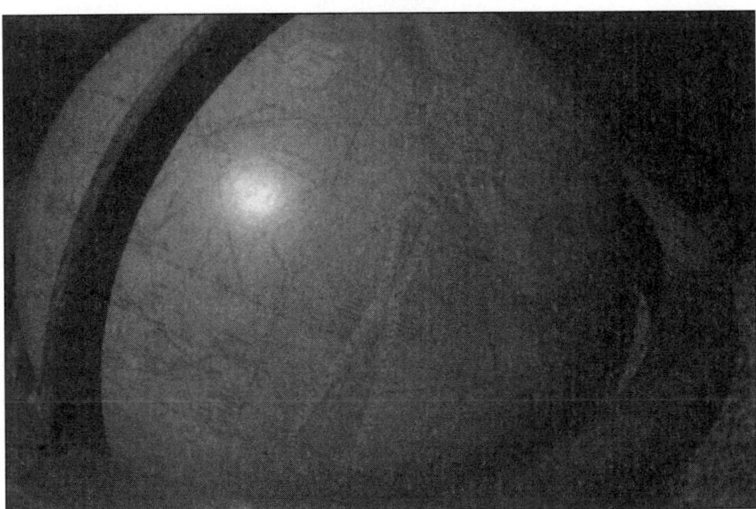

FIGURE 26-1. A pre-World War I glass globe clearly shows the figure eight shape known as the analemma, which is used to determine the position of the sun in the sky from anywhere on Earth. Most globes no longer show it.

FIGURE 26-2. As shown, taking a fixed camera photo of the sun every day of the year produces the analemma. Photographer Frank Zullo combined thirty-seven separate images of the sun over Large Sundial, Carefree, Arizona, to show this unique phenomenon.

On a globe or map, find Earth's equator and imagine extending it into the sky, where it becomes the **celestial equator** on the **celestial sphere** (dome of stars). If it were possible to photograph the sun and celestial equator in the sky at the same time each day, you would notice the sun's place relative to the celestial equator changes constantly. The sun's annual path in the sky, called the **ecliptic**, crosses the celestial equator at only two points, called **nodes**. The angle between the ecliptic and celestial equator is about 23.5°, the same tilt that Earth's axis has compared to its orbital plane (Figures 27-1 and 26-4).

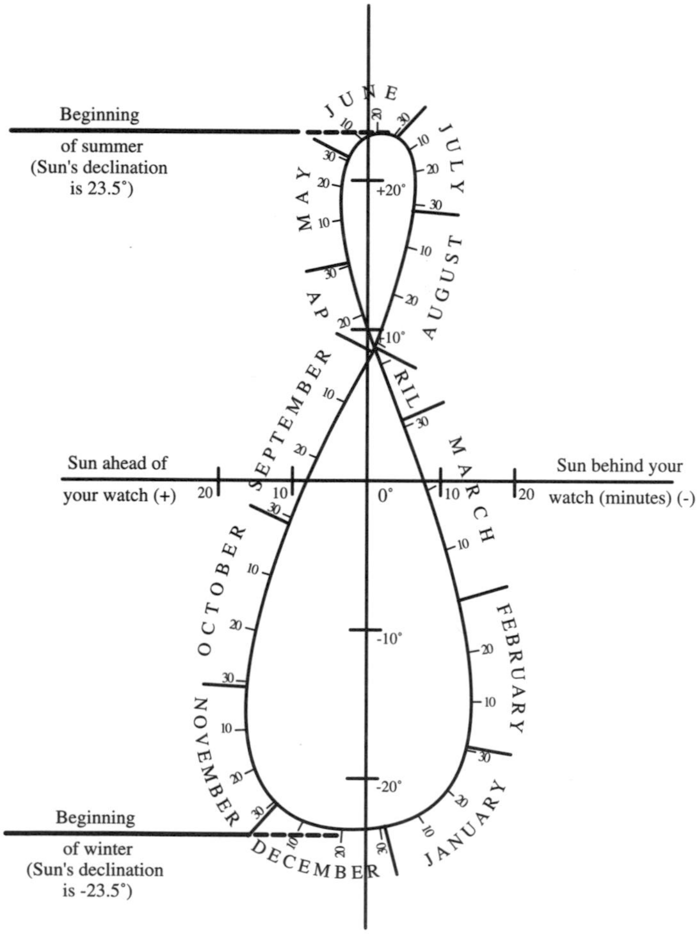

Each day on the analemma gives two pieces of information, the first being the height (or altitude) of the sun relative to the celestial equator, called **declination**. For example, reading from the analemma in Figure 26-3, on or about December 21 the sun's declination is about –23.5° (its southernmost point during the year) because Earth's northern hemisphere is tilted **away** from the sun. In contrast, around June 21, the sun's declination is about +23.5° (its northernmost point), since Earth's northern hemisphere is tilted **toward** the sun. The north-south span of the analemma figure, a direct result of the tilt of Earth's axis, covers about 47°, the difference in altitude of the sun on the first day of summer and the first day of winter.

To interpret the east-west shape of the analemma, we need to jump a little into the world of astronomical timekeeping. In keeping time by the sun, it is important to remember that the sun crosses the sky each day as a

FIGURE 26-3. The analemma, a reverse of the global analemma. (Why? See Figure 26-4.)

result of Earth's rotation. The sun's annual motion is more complex and less noticeable. If it were possible to sketch the sun each day against the background constellations, you would notice that the sun gradually moves **eastward** along the ecliptic during the course of a year, **though not at a constant rate**. To understand this, walk alternately slowly and quickly around a large tree, watching the tree trunk while noticing objects in the background. At different points in your "orbit" you will see the trunk move at different rates against different background objects. The same occurs with the sun as we orbit it, except the background objects are very distant stars.

It might then be natural to ask, "When is the sun moving fastest and slowest?" The annual movement of the sun is really a reflection of how quickly Earth orbits the sun, so in early July, when we are farthest from the sun and moving slowest, the sun moves its **slowest** against the stars of the Zodiac. In contrast, in early January, Earth is closest to the sun and moves its fastest, causing the sun to appear to move its **fastest**. This motion refers to the **True Sun**, since it is a reflection of how the sun is actually seen to move during the year. But following True Sun time is difficult, so astronomers have created a simpler timekeeping device based on a sun that would not travel fast or slow but always at the same rate. Called the **Average Sun** or **Mean Sun**, this is the same as the clock time each of us reads from a watch and represents the sun travelling a **constant** 1 degree per day eastward along the ecliptic.

Returning to Figure 26-3, notice that the horizontal line passing through September and March is marked in minutes, with "Sun Ahead" appearing on the September side and "Sun Behind" appearing on the March side. This represents how far ahead or behind (in minutes) the True Sun is from our clock time, the Mean Sun. Usually, "noon" is judged to be the moment when the sun is exactly halfway across the sky. But the sun is at this point when your watch reads 12 noon on only four days of the year (April 15, September 3, June 15, and December 25). At other times of the year, the sun is either running "slow" or running "fast" relative to the halfway point in the sky, called the **meridian** (Figure 26-4). Thus if your

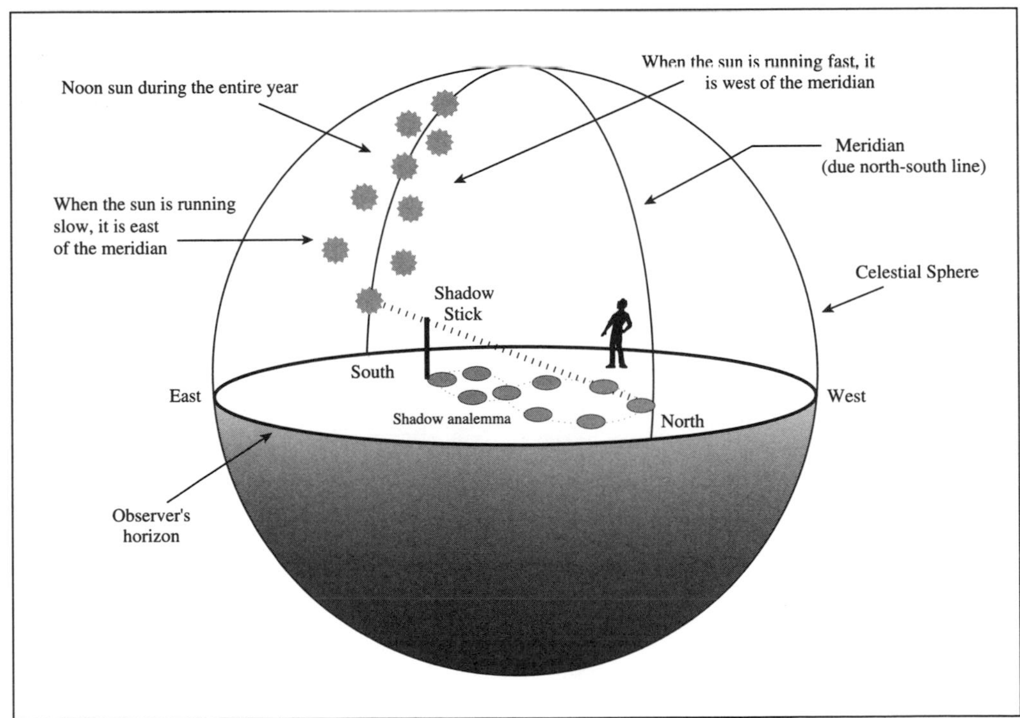

FIGURE 26-4. The different positions of the noon sun, (relative to the meridian), during the year trace out the analemma (note the shadows are a REVERSE of the actual sun positions).

watch reads 12 noon and the sun is behind clock time, the sun will appear slightly **east** of the meridian (January, February, March, July, and August). When it is ahead of clock time, the noon sun appears just **west** of the meridian (May, September, October, November, and most of December).

The minus (–) times in Table 26-1 represent when the sun is running slow (behind clock time) and the other times are when the sun is running fast. For example, February 14 is a minus day; the True Sun is fourteen minutes **behind** your clock, so when your watch reads noon, it will still take the sun fourteen minutes to reach the halfway point in the sky. But May 11 is a plus day; the True Sun is four minutes **ahead** of your clock, so when your watch reads 12 noon, the sun already passed the halfway point four minutes ago.

So now we know what causes the shape of the analemma. The vertical span of the figure is a direct result of the tilt of Earth relative to its orbit. As we orbit it, the sun takes a high position in the sky during summer (the top of the analemma) and a low position in the winter (the bottom of the analemma). The horizontal span of the figure is a combination of a portion of the apparent sun's movement at the equinoxes and the solstices, a less direct effect of Earth's tilt. The actual sun's movement is different at the equinoxes than at the solstices. Lastly, Earth's elliptical (or eccentric) orbit contributes to the horizontal span of the figure. During winter, when Earth is closest to the sun, the sun moves faster along the ecliptic (giving the wide bottom to the figure-eight), whereas in summer, when we are farthest from the sun, the sun moves slower along the ecliptic (giving the narrow top of the figure-eight).

ACTIVITY 1

In this activity you will be able to periodically mark off shadows cast by a long stake. After a year, these markings will connect to form the characteristic figure-eight of the analemma. As shown in Figure 26-5, drill two ¾-inch-diameter holes in the plywood, marking one end "south" and the other "north." Sand and clean each hole until the dowels can slide easily and loosely through the holes. Find a level clearing in your yard that is free from activity and has a year-round unobstructed view of the southern sky. Drive the long dowel into the ground until about 12 inches remain protruding.

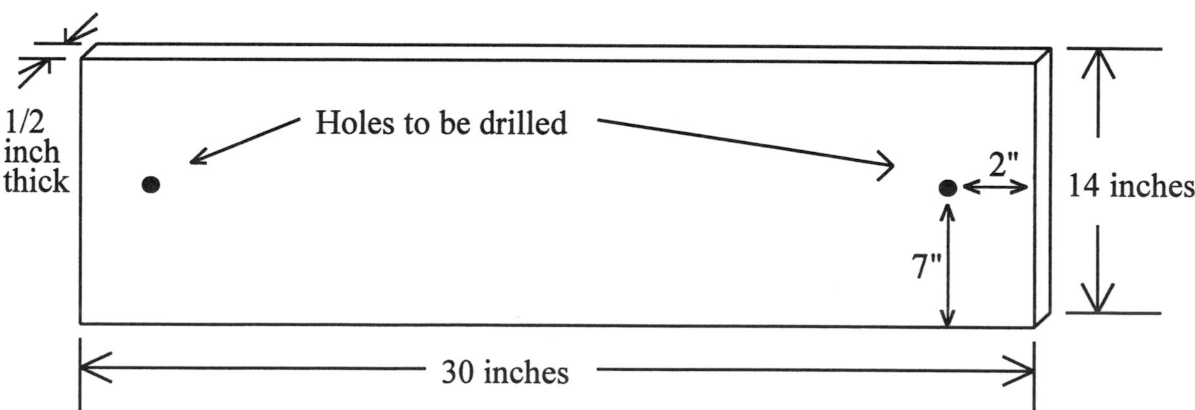

FIGURE 26-5. As shown, drill ¾-inch holes centered 2 inches from each end and 7 inches in from each side.

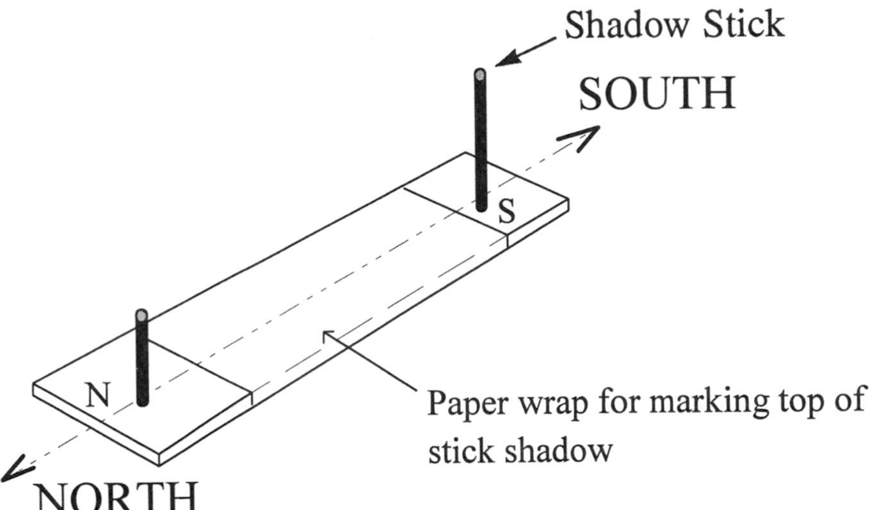

FIGURE 26-6. The taller of the dowels is the shadow stick and must be on the southern side of the board, while the shorter stick just serves as a reference mark for proper north-south placement of the board.

Next, you need to find due south (Figure 26-6). One precise method is to place the south end of the board over the stick, with the board flat on the ground, on April 15, September 3, June 15, or December 25, at **precisely** noon. It is important to use only these dates since they are the only times during the year when the noon sun is due south. Tie the string around the stick and carefully center it along the stick's shadow, moving the board until the second hole meets the string. Drive the second stake in here. (A less-accurate method of finding due south is by sighting the North Star; the direction opposite is south.) Now lift the board off and wrap it with the paper, taping it on the back side.

Three or four times a month, go out just before noon and slide the board down over the stakes (south end over the larger stake). Then, at **exactly** noon standard time (1 P.M. during daylight saving time), mark and date the center of the tip of the shadow cast on the paper by the south stake (the shadow will look fuzzy). Take the board indoors; the dowels remain outdoors. After a year's worth of shadow marks, connect the marks to produce the analemma.

ACTIVITY 2

For those of you who like to draw and would rather not devote a full year to discovering the analemma, you can graph the position of the sun for an entire year. First, choose a date, say February 14. Refer to Table 26-1 to find the sun's declination for that date, and how far behind or ahead of your watch it is. For February 14, the sun's declination is –13° and it is 14 minutes **behind** your watch. Next, on the graph, find –14 on the time line and –13° on the declination line. Lightly draw a straight line through both points until they intersect. Make a dot at their intersection and mark the date "2/14"; then erase the lines you drew. Do this for at least thirty-five dates and then connect the dots. Congratulations! You've made an analemma.

Note: The dates on this graph are a mirror image of those in Activity 1 because the shadow analemma is a reversal of the analemma in the sky (see Figure 26-4).

TABLE 26-1

This table gives values of how far ahead or behind the sun is relative to your watch (in the minutes column) as well as the sun's declination for given dates. By plotting the minutes on the horizontal and the declination on the vertical, you will get the analemma (see Figure 26-3).

DATE	MINUTES THE SUN IS AHEAD OR BEHIND (-)	DECLINATION OF SUN
January 1	–3	–23°
January 10	–7	–22°
January 25	–12	–19°
January 30	–13	–18°
February 4	–14	–16°
February 14	–14	–13°
February 24	–13	–10°
March 1	–12	–8°
March 11	–10	–4°
March 21	–7	0°
March 31	–4	4°
April 5	–3	6°
April 15	0	10°
April 25	2	13°
May 1	3	15°
May 11	4	18°
May 21	3	20°
June 1	2	22°
June 11	1	23°
June 22	–2	23.5°
July 7	–5	23°
July 13	–6	22°
July 23	–6	20°
August 2	–6	18°
August 13	–5	15°
August 28	–1	10°
September 3	0	8°
September 18	6	2°
September 28	9	–2°
October 8	12	–6°
October 19	15	–10°
October 29	16	–13°
November 8	16	–16°
November 22	14	–20°
December 7	9	–23°
December 22	2	–23.5°

Graph of Declination of Sun vs. Time

(AHEAD OR BEHIND NOON)

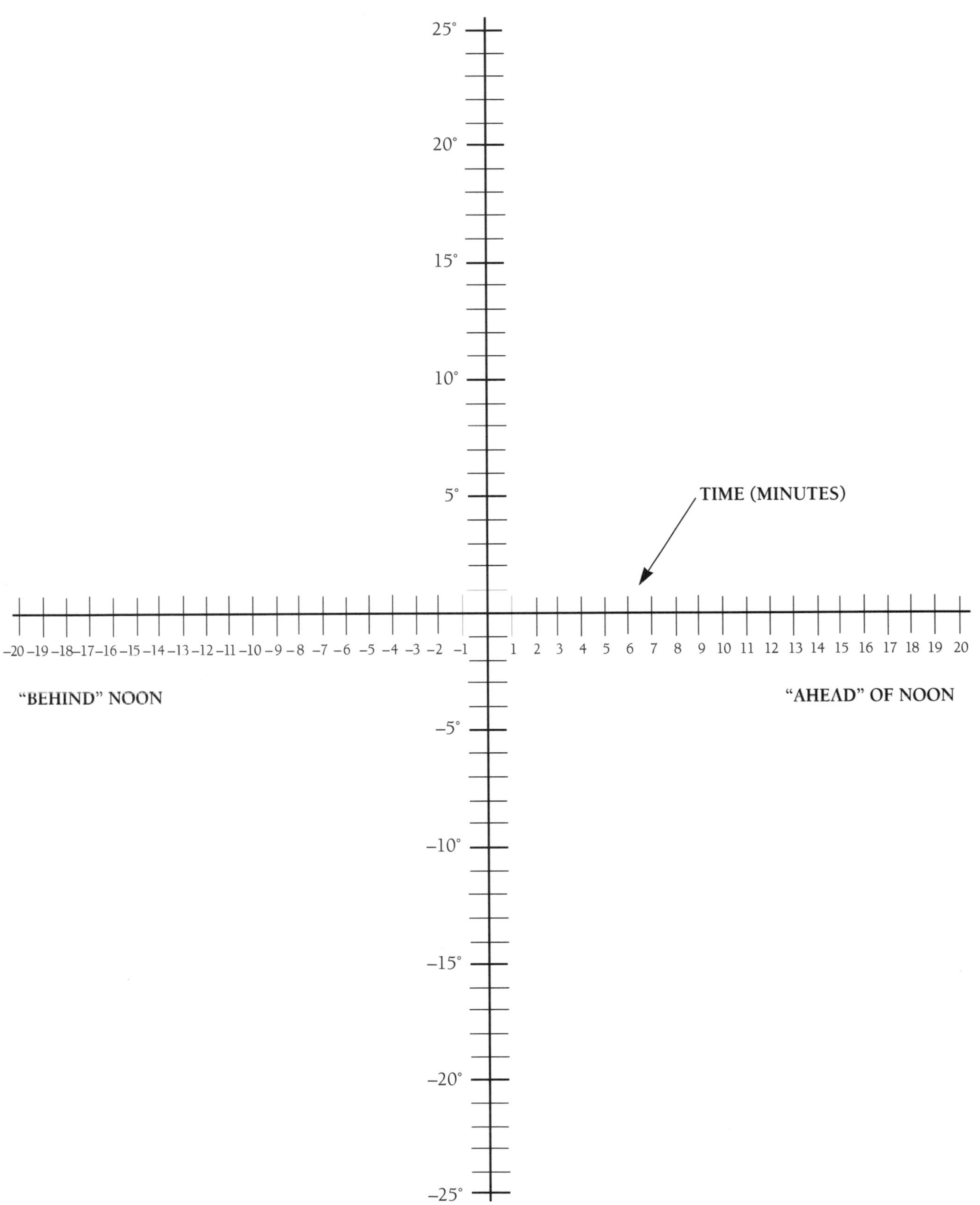

27 Chasing Solar Eclipses

Level: All

Objective: To capture the excitement of a solar eclipse

Materials:
- Mylar sun viewers
- A date when the sun is in eclipse (see Table 27-1)
- Thermometer, light meter, watch, pad and pencil

BACKGROUND

What first appeared as a tiny bite on the sun's edge grows into a large scoop of darkness, ominously engulfing the sun. After about two thirds of the sun has been swallowed by the encroaching blackness, the air around you takes on a bizarre deep, dark, steel-gray color. You feel the wind kick up, birds quiet, insects sing, and animals disappear with the apparent onset of night. Before your eyes, nearly the entire sun is swallowed by the trespassing blackness. Just moments before the sun vanishes completely, a final surge of sunlight falls upon you as the sun is transformed into a grand diamond ring in the sky. Now, where the sun **was**, you see a perfect, round black void amidst the brightest stars and an entire horizon of eerie orange. The next few minutes pass in what seems like a fraction of a second, with the sun returning quickly. It is over. Astounded at this spectacle, you ask, "When can I see this again?"

This is a total eclipse of the sun, inspiring both excitement and awe. Viewed by the ancients as a cause of fear, such an event now attracts people from thousands of miles away, carrying telescopes, cameras, and binoculars. There is nothing to fear, as an eclipse of the sun is simply the darkness caused by passage of the moon between Earth and sun, preventing sunlight from reaching part of Earth. Though care must be taken in observing the sun during any solar eclipse, the event is always bound to bring exhilaration and excitement.

For a solar eclipse to occur, the moon must be in its "new" phase, when it lies between the sun and Earth. A glance at Appendix 3 listing the lunar phases shows that there are some twelve to thirteen new phases each year, though no more than two or three solar eclipses in the same time. Why? As described in Activity 21, the path of the moon around Earth is tipped a little relative to the ecliptic, the sun's path in the sky. Only when both of these paths cross (a point called a **node**) at new moon will a solar eclipse occur (Figure 27-1).

There are three types of solar eclipses. If the moon's orbit at new moon does not exactly coincide with the sun's path in the sky, a **partial solar eclipse** can occur. Only part of the sun is blocked by the moon. Usually, no noticeable darkening occurs unless about 75 percent of the sun is covered. The temptation at this point is to look up to see what's wrong with the sun, but remember, DO NOT OBSERVE A PARTIALLY ECLIPSED SUN WITH THE UNPROTECTED EYE! Instead, use a **mylar sun filter** or a **projected** solar image (see Activity 22). Solar filters are designed to reduce the sun's intensity while absorbing ultraviolet and infrared radiation that could otherwise burn your eye's retinas. If you use a projected image of the sun, try to watch the moon's edge cover and uncover sunspots.

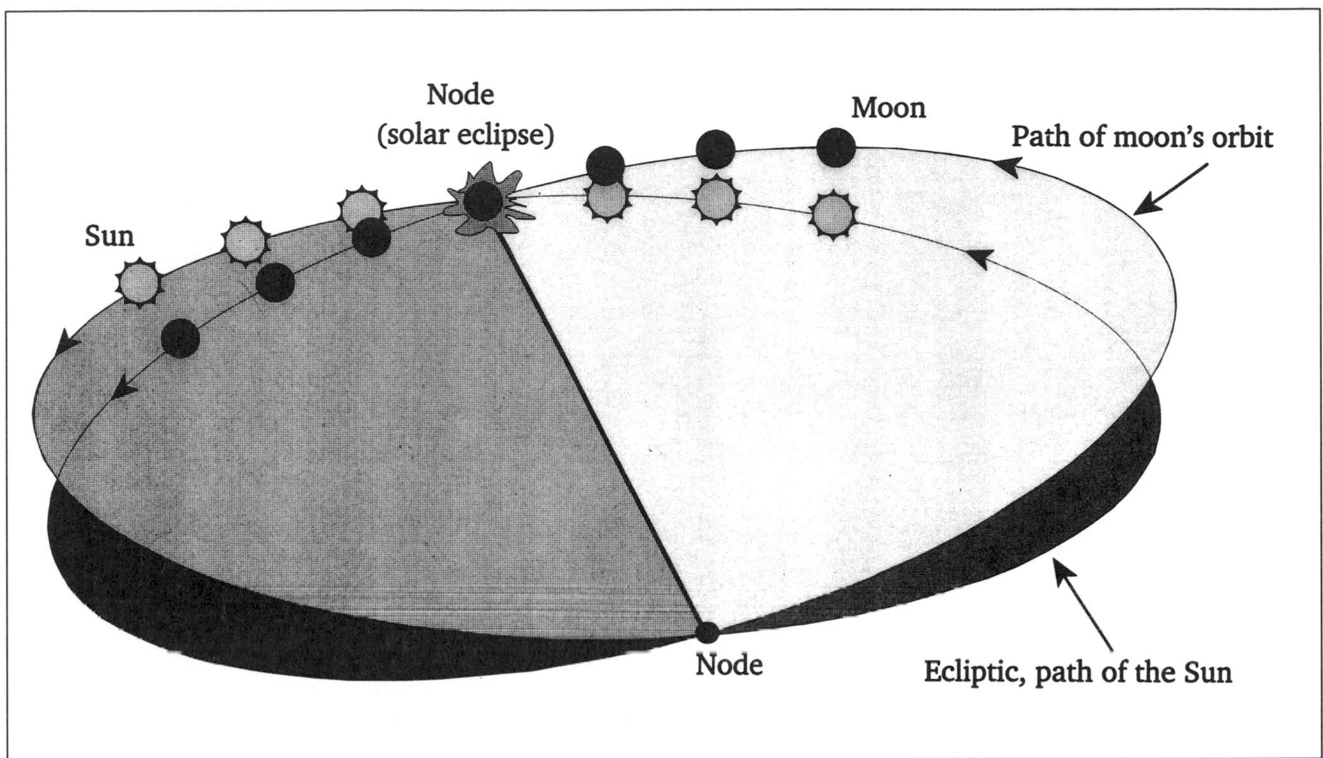

FIGURE 27-1. The paths of the sun and moon shown together demonstrate why eclipses occur only when the sun and moon meet at a point common to both paths (called a node). Notice that the moon's orbit is tilted by about 5° relative to the ecliptic.

More exciting is an **annular solar eclipse**, the word "annular" being derived from the Greek word "annulus," meaning "ring." At maximum eclipse, the moon blocks all but a ring of sunlight in the sky (Figure 27-2). The sun and the moon usually appear to us to be the same size in our sky. But both Earth's orbit about the sun and the moon's orbit about Earth are elliptical (not circular), causing the Earth-sun and Earth-moon distances to vary considerably. An annular eclipse occurs at times when the new moon is at a node but appears too small to cover the sun's disk completely, thus leaving a ring of sunlight visible. Again, a solar filter or projection must be used to view this type of eclipse.

FIGURE 27-2. During an annular eclipse, the sun appears as an "annulus," or ring.

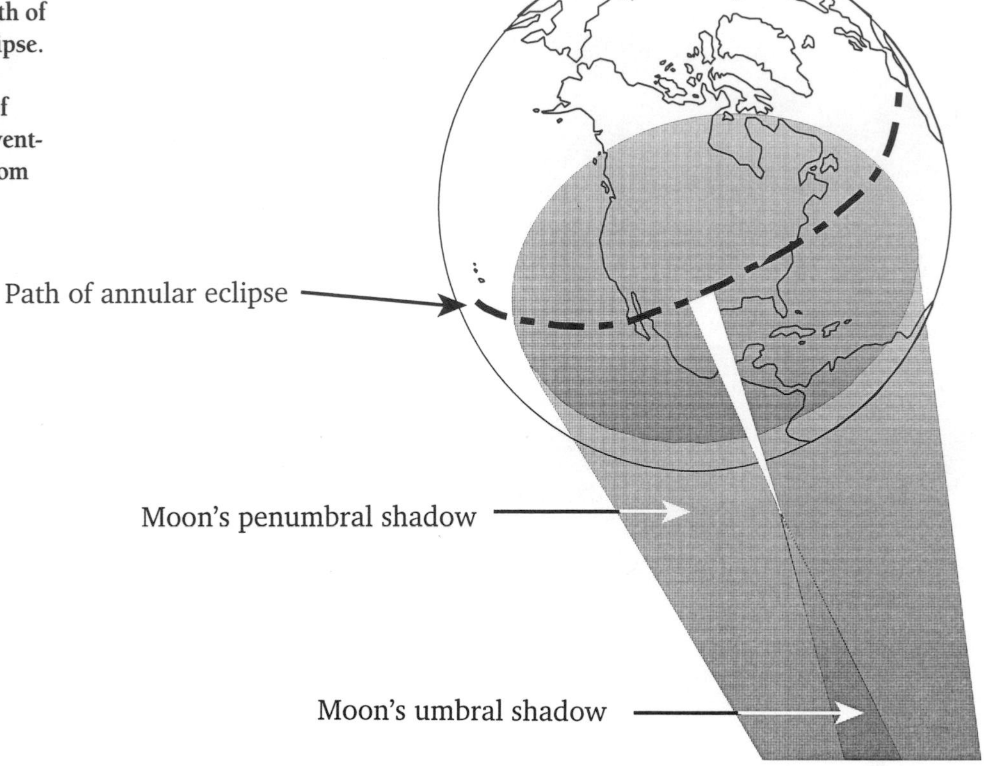

FIGURE 27-3. The path of an annular solar eclipse. The moon's umbral shadow falls short of Earth's surface, preventing a total eclipse from being seen.

The most spectacular eclipse is the breathtaking **total solar eclipse**, an event to which few things in nature can compare. The moon's angular size is equal to or greater than that of the solar disk, allowing the entire sun to be covered. **First contact** marks the very beginning of the moon's passage in front of the sun (Figure 27-4a). The interval from **second** to **third contact** is when the sun is completely covered, the period called **totality**. Moments before you are immersed in the moon's shadow, the last burst of sunlight from behind the moon creates the **diamond ring effect** (Figure 27- 4b). During totality, you may see pinkish-red flames of hydrogen gas pouring off the sun's surface as well as the sun's crown (the **corona**) spraying radially outward from the sun (Figure 27-5). Since no part of the brilliant sun is visible, the period of totality can be safely observed without any filters—but filters *must* be used before and after. Lasting no more than about seven minutes, totality seems to fly by. The eclipse officially ends at **fourth contact**, with the moon's disk moving off the sun. The entire event, from first through fourth contact, can last about three hours.

Long-term and highly detailed eclipse maps are available from NASA (see Appendix 2), to help viewers pinpoint exact locations on the globe from which a particular eclipse will be visible. Unlike lunar eclipses, which are visible over the entire night side of Earth, solar eclipses are visible only over very narrow stretches of Earth's surface. During a total solar eclipse, the moon's shadow on Earth may be no more than 100 miles wide (Figure 27-6). Some eclipses are visible only from oceans—in which case, solar eclipse fanatics will gather in a jetliner to fly through the path of the moon's shadow or board a ship to see the event.

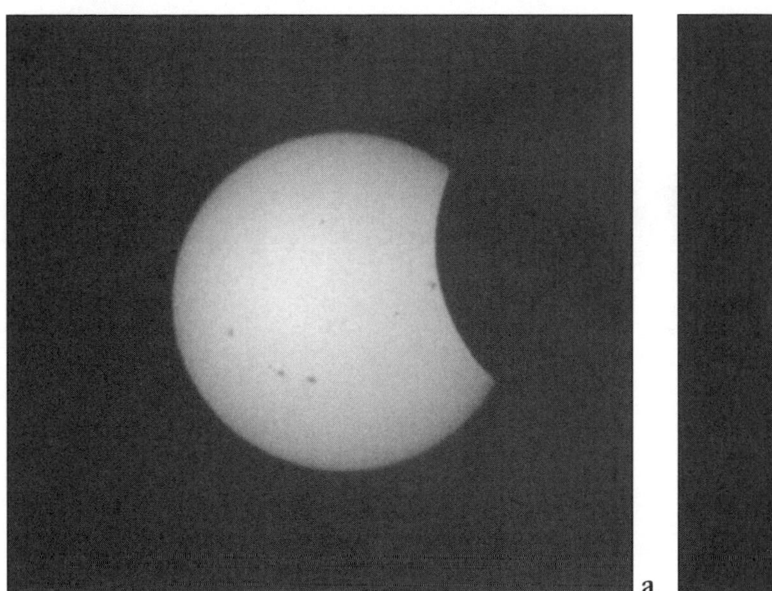

a

b

FIGURE 27-4. The early stages of a solar eclipse begin with the moon gradually covering the sun, (a). As the last bit of sun light streams by the moon's edge, this burst and the corona of the sun creates the Diamond Ring effect, (b).

FIGURE 27-5. As shown, the moon completely covers the sun during a total solar eclipse (seen here from Clarkleigh, Manitoba, in February 1979). The sun's corona is fully visible. (Photo courtesy of Vanderbilt Planetarium)

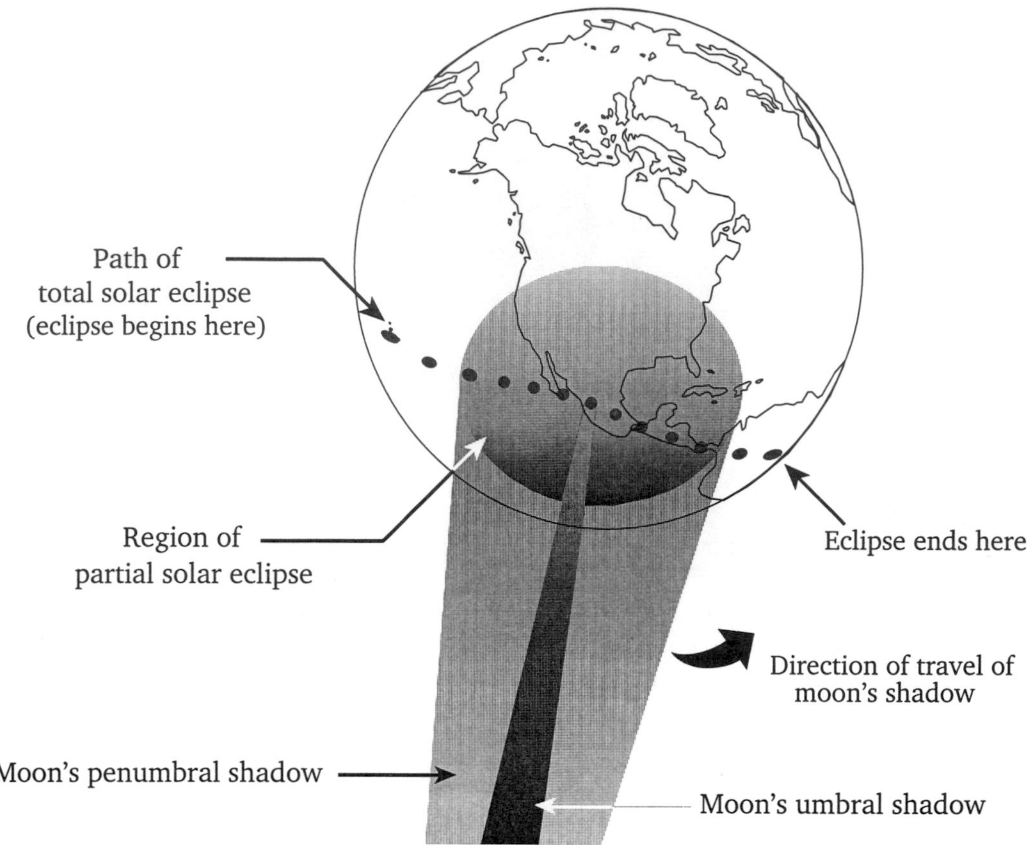

FIGURE 27-6. Path of the 1991 total solar eclipse across Earth. People within the moon's umbral shadow see the sun completely covered, while those in the penumbral shadow see it partially covered.

ACTIVITY

Seeing any type of solar eclipse is an inspiring and enjoyable experience, especially the first time. You don't need thousands of dollars worth of telescope equipment and camera gear to enjoy one. Here are some ideas for things to do when you have the good fortune to witness a solar eclipse:

- Use a watch to time the event from the very beginning to the very end of the moon's passage across the sun's disk (be it a partial, annular, or total eclipse).
- While part of the sun is still visible, look at the images of the sun on the ground formed by the leaves of trees. (Gaps between the leaves act as the "lens" on a kind of pinhole camera, forming small "crescent sun" images on the ground.)
- If you have a projected or direct filtered telescopic view of the sun, count the number of sunspots.
- As it darkens outdoors, notice the **color** of this darkness. Does it have any odd characteristics?
- Take light-meter readings, using an actual light meter or a camera containing one. Try taking photos before, during, and after the eclipse at the same shutter speed. What do they look like?

- Is it dark enough to see stars and planets? If so, which ones can you see? What is the dimmest star that you can detect?
- Take temperature and wind readings as darkness falls. Wind readings can be done by "feel" rather than with a device, but get an outdoor thermometer and read the temperature at intervals of perhaps ten to twenty minutes.
- Observe nature. What do you notice about birds and animals before, during, and after the eclipse?

As you watch the eclipse, you will surely discover the fascination that people have always felt for this beautiful event. Soon you will ask, "When can I see this again?"

TABLE 27-1

Solar Eclipses, 1994–2000

DATE	TYPE OF SOLAR ECLIPSE	MAXIMUM DURATION (MINUTES: SECONDS)	REGION OF VISIBILITY
May 10, 1994	Annular	6:13	Eastern Pacific Ocean, central U.S., North Atlantic
November 3, 1994	Total	4:24	Central S. America (Peru, Chile, Paraguay, Brazil)
April 29, 1995	Annular	6:38	Northern S. America
October 24, 1995	Total	2:10	Southeast Asia (Iran, Burma, Vietnam, Pacific Ocean)
April 17, 1996	Partial	—	South Pole, Antarctic region
October 12, 1996	Partial	—	Greenland, Northern Europe, North Pole
March 9, 1997	Total	2:51	Northern Russia, Arctic Circle, Mongolia
September 1, 1997	Partial	—	Australia, New Zealand, Antarctic Circle
February 26, 1998	Total	4:09	Pacific Ocean, northern S. America, Aruba, Caribbean, western Africa
August 22, 1998	Annular	3:14	Malaysia, Indonesia, S. Pacific
February 16, 1999	Annular	1:19	Indian Ocean, Australia
August 11, 1999	Total	2:23	Central Europe (England, France, Germany), Iran, India
February 5, 2000	Partial	—	Antarctica
July 1, 2000	Partial	—	S. Atlantic, southern S. America
July 31, 2000	Partial	—	North polar region
December 25, 2000	Partial	—	Northern Central America, U.S.

SECTION IV

The Solar System

28 Hide and Seek

Finding Elusive Mercury

Level: All

Objective: To observe the planet Mercury

Materials:

- Binoculars or a small telescope
- An unobstructed view to the west or east
- One of the following periodicals:
  Sky & Telescope
  Astronomy
  Old Farmer's Almanac
  Observer's Handbook
  Sky Calendar
  Astronomical Calendar

BACKGROUND

Orbiting the sun only 36 million miles from its fiery surface is tiny Mercury, second smallest planet in the solar system. Mercury scoots around the sun once every eighty-eight Earth days, the fastest orbit of any planet. Close-up photographs taken by the passing eye of the Mariner 10 spacecraft in 1974 revealed Mercury as a barren, airless, waterless world whose heavily cratered surface shows the scarring effect of meteorite bombardment. When fully lit by the sun, the rutted surface of Mercury broils at a staggering 800° F. In the cool of the Mercurian night, however, the planet's surface temperature dips to a frigid –200° F! No other planet experiences such a stark temperature change.

ACTIVITY

Because Mercury is held captive so close to the sun, it is never visible in a fully darkened sky. Mercury is visible only in bright twilight, either very low in the western sky immediately after sunset or very low in the eastern sky just before sunrise. It's not surprising that many stargazers can spend years enjoying the night sky without ever having seen this elusive planet. The purpose of this activity is to help you join the elite group of "Mercury watchers."

First determine when Mercury will be visible in the sky by consulting one of the publications in the materials list at the beginning of this activity. Since it takes Mercury less than three months to complete an entire orbit, its visibility changes rapidly. Within that three months, it will be visible in the evening sky, swing in front of the sun, pop out in the morning sky, swing behind the sun, then return back to the evening sky.

Whenever Mercury (or Venus) passes near the sun, astronomers say that the planet is in **conjunction** with the sun. For instance, every time Mercury shifts from our evening to our morning sky, it passes between Earth and sun (not necessarily directly in front of the sun, however), referred to as **inferior conjunction.** Likewise, Mercury passes behind the sun, or through **superior conjunction**, when it goes from the morning to the evening sky. Figure 28-1 illustrates what is going on.

FIGURE 28-1. The planet Mercury's orbit of the sun as seen from Earth. Note the points of inferior conjunction and superior conjunction.

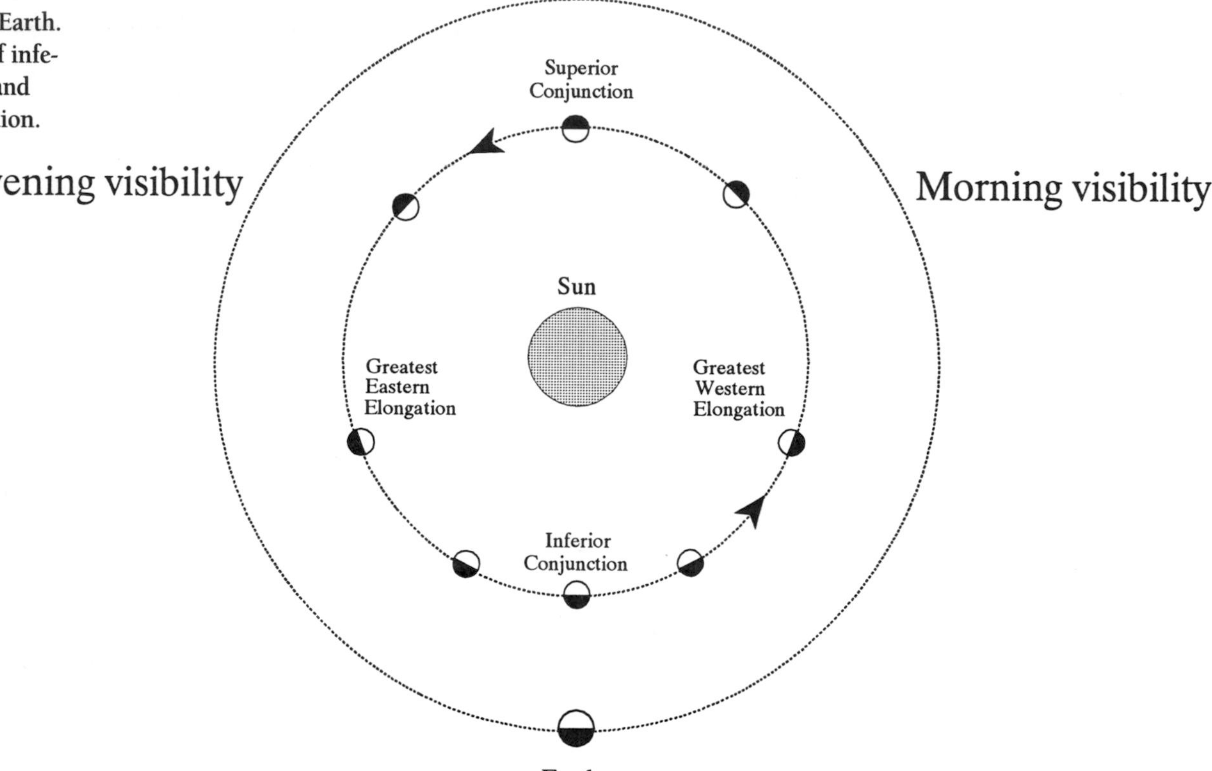

Since the window of visibility is so short, observers must know exactly when and where to look for Mercury, or they will be unsuccessful. The best time to look for Mercury is when it is at **greatest elongation** (that is, maximum distance from the sun in our sky). Greatest eastern elongation (when Mercury is farthest east of the sun) places the planet in the western sky of early evening. Mercury reaches greatest western elongation (when the planet is farthest west of the sun) when it is visible in the early morning eastern sky.

For most of us, it is much more convenient to look for Mercury when it is at or near greatest eastern elongation in the early evening western sky. But some eastern elongations are better than others. Due to the angle the sun's path makes with our horizon, the late winter and early spring are usually the best times for residents of the northern hemisphere to spot Mercury in the evening sky. For early birds, the late summer and fall are best for Mercury spotting when it is at greatest western elongation in the early morning sky.

Even at greatest elongation, Mercury may still elude your search. To spot the planet, find an open observing site with a good view in the direction of sunset (or sunrise). The best places for Mercury-spotting (and stargazing in general) include a beach, a large field, or the top of a hill. A hilltop was used to capture the naked-eye view of Mercury, along with three other planets and the moon, shown in Figure 28-2.

Figure 28-2. Mercury is but one of four planets in this photograph. On the bottom are Mercury (right) and Jupiter (left), in the middle is Mars (right) and the star Regulus (left). Finally, in between Regulus and the crescent moon is brilliant Venus. (Photo by Brian Kennedy)

Binoculars are a great help when looking for Mercury. About half an hour before sunrise or after sunset, scan the horizon near the sun-point very slowly. Look for a fairly bright point of light shimmering in the twilight glow. A note of caution: Mercury is easy to mistake for a bright star. Only by knowing its exact location can a positive identification be made.

Once Mercury is spotted, study it with a telescope. Can you make out any shape? Due to air turbulence in our atmosphere, Mercury may look like a "boiling" speck. But if the atmosphere is steady, it might be possible to see Mercury exhibiting phases like the moon. Little additional detail will be visible.

Mercury is much harder to find than any of the other naked-eye planets. But with a little planning, you'll be able to have a good look at elusive Mercury.

29 Following the Phases of Venus

Level: All

Objective: To observe the planet Venus going through phases

Materials:

- Binoculars or a small telescope
- Photocopies of Venus
- Observation Form (included with this activity)
- One of the following periodicals:
  Sky & Telescope
  Astronomy
  Old Farmer's Almanac
  Observer's Handbook
  Sky Calendar
  Astronomical Calendar

CLOUDY WITH A CHANCE OF RAIN
The weather forecast for Venus is always cloudy, always hot, with a good chance for showers. Its surface is perpetually shrouded in an impenetrable atmosphere of poisonous carbon dioxide. Deep beneath the clouds, the surface of Venus is the hottest of any planet, with an average temperature of about 900°F! And when it rains on Venus, it doesn't rain water; it rains SULFURIC ACID.

BACKGROUND

Venus, at times the closest planet to Earth, appears like a dazzling diamond whenever it is visible. Like Mercury, Venus has an orbit that is closer to the sun than our own, restricting the planet's visibility to early evening or early morning. But while Mercury's appearance in the sky is tentative at best, there is no mistaking Venus.

Venus and Earth are frequently called the twin planets because of their similar diameters and masses. Earth measures about 8,000 miles across to Venus's 7,600-mile diameter. But the similarity ends there, for Venus is a hostile world, not at all like the planet we call home.

ACTIVITY

Venus reveals very few of its secrets when viewed from our safe haven here on Earth. All that telescopes reveal of our "twin" is a pearl-white disk that appears to go through phases similar to our own moon's.

It is always exciting to track these phases as Venus swings around the sun. At times, Venus looks like a tiny gibbous moon, displaying a nearly circular, though small, disk. Other times, Venus shows a half disk similar to the first-quarter or last-quarter moon. Still other times show Venus to have grown in apparent size, while also changing to a crescent phase.

To find out where, or if, Venus will be in tonight's sky, consult Appendix 4 at the end of this book or one of the publications in the materials list at the beginning of this activity. If it's in the evening sky, you can't miss it.

If you have a telescope or binoculars, take a look at Venus. What phase is Venus in? Do you see any color on the planet? (Don't confuse the planet's color with bluish fringes around its edges. They are caused by an optical imperfection called **chromatic aberration**, common to less-expensive refracting telescopes.)

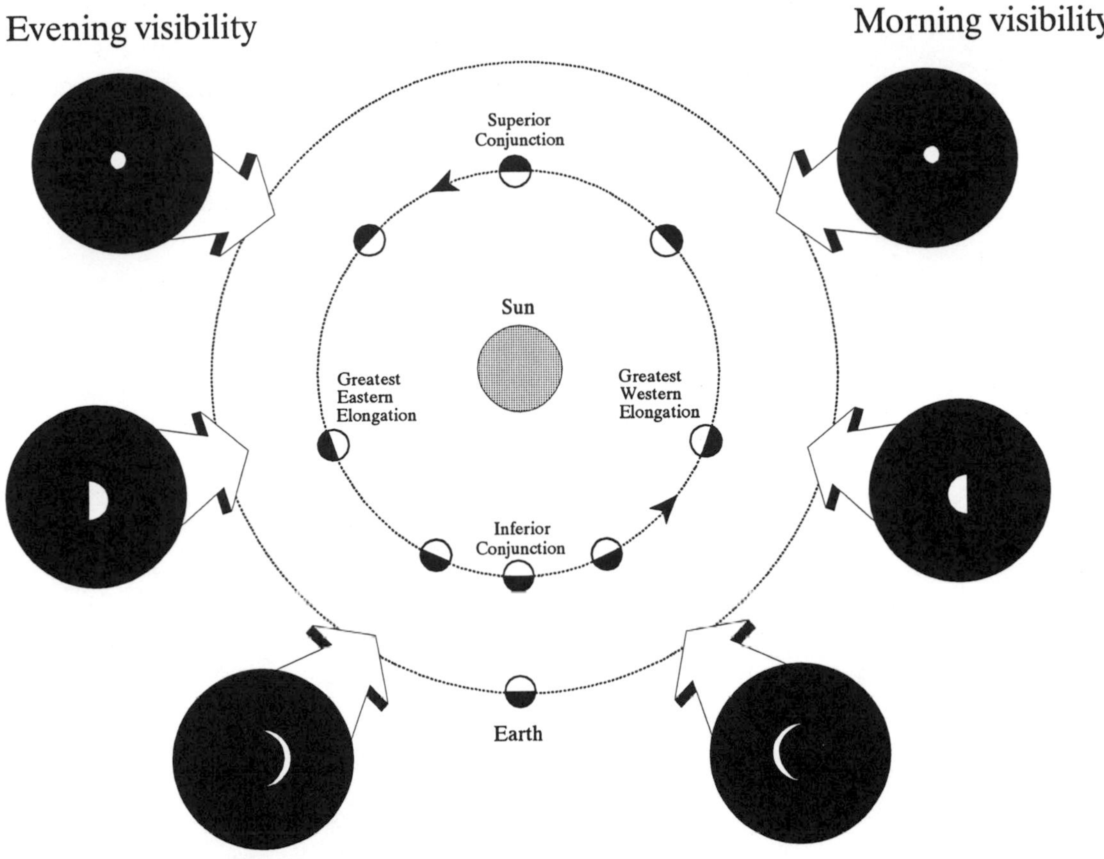

FIGURE 29-1. (above) The orbit of Venus around the sun as seen from Earth. Shown at each point in the orbit is the appearance of Venus as seen through an Earth-based telescope.

FIGURE 29-2. (left) Venus as seen through a small telescope.

By watching the planet's phase, an observer can easily determine where Venus is in its orbit. Figure 29-1 will help. When Venus appears as a crescent (Figure 29-2), we know it is near the Earth, close to its inferior conjunction point between our planet and the sun. At greatest elongation, Venus appears as a wide crescent or about half lit. When it is farthest from Earth, near superior conjunction, Venus appears similar in shape to a gibbous moon. (For definitions of elongation, and inferior and superior conjunction, see the discussion in

Venus Observation

Date: ______________________ Time: ______________________

Telescope: ______________________ Power: ______________________

Notes: ____

____

____

____

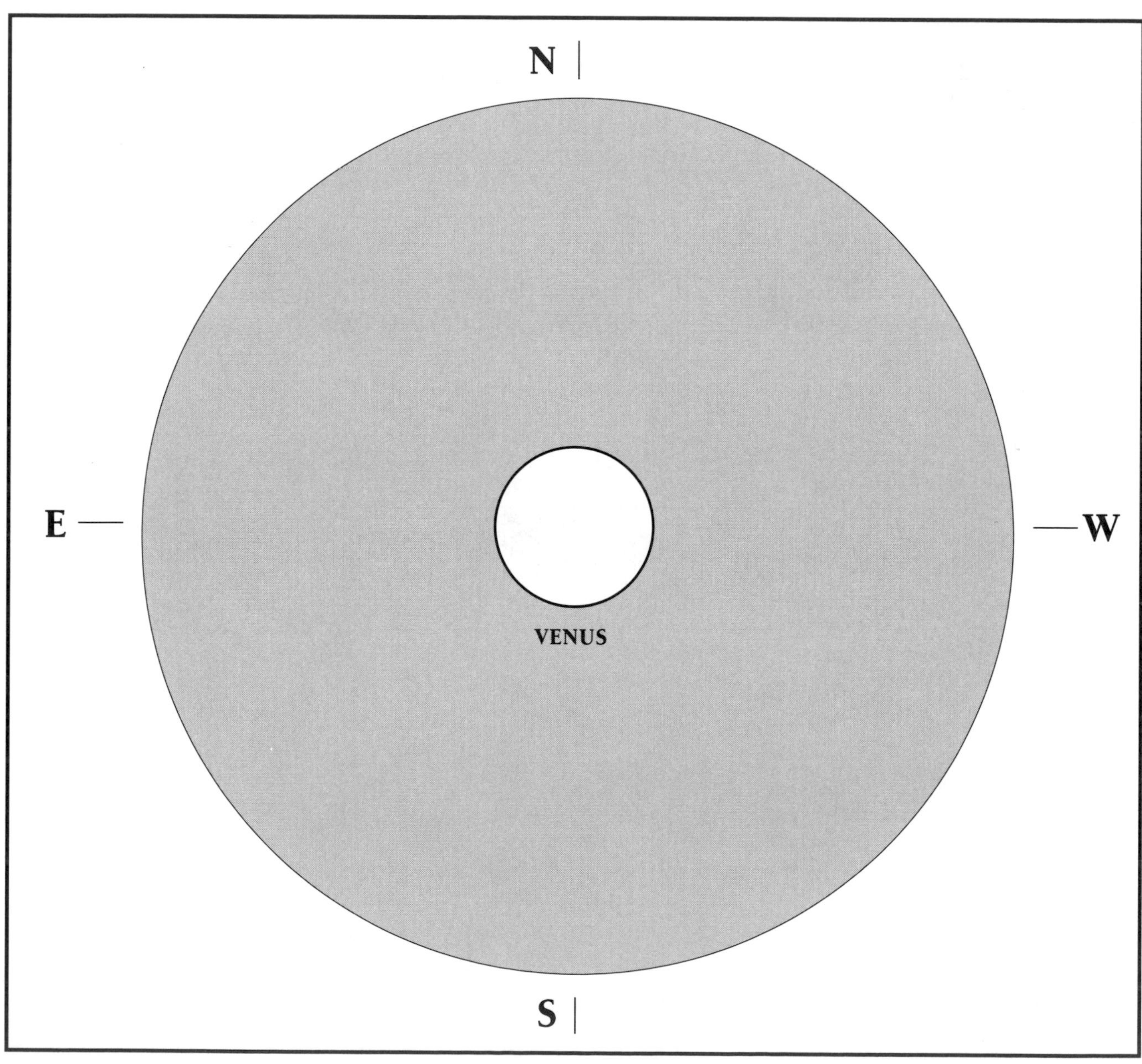

Activity 28 about Mercury.) Keep an ongoing observing log by recording your impression on a Venus Observation Form every time you look at the planet.

By tracking the phases of Venus, we can follow in the footsteps of Galileo, who became the first person to view and describe the phases in the early seventeenth century. Before his historic observation, it was believed that Earth was at the center of the solar system (indeed, at the center of the universe). But Galileo's observations showed that Venus went through phases and, more importantly, that it appeared to change in size as it phased, proving that it orbited the sun, not Earth. (Venus could still go through phases if both it and the sun circled Earth, but it would not change in apparent size as it does.)

The next time Venus is visible in the sky, take a good look at our "twin" planet—but remember just how unlike Earth our sibling actually is.

30 The "Dizzy" Planet

Following the Backward Motion of Mars

Level: Intermediate and advanced

Objective: To learn about the "forward" and "backward" movement of planets among the stars, with emphasis on Mars

Materials:

- A big yard or park
- A friend
- Tracing paper, red pencil, regular lead pencil

BACKGROUND

Telescopic observations of Mars in the mid-1600s revealed a world with polar caps, dark regions, and a variety of changing surface features. During the late nineteenth century, observation revealed a series of crisscross patterns, mistaken for canals. This observation, along with the desire to discover life beyond Earth, sparked the idea that living beings must inhabit this seemingly Earth-like planet. Thus Mars has been the subject of intense study for the entire twentieth century. To the disappointment of some, the U.S. Viking I and II landers in 1976 showed that Mars holds no known life or canals.

Named for the Greek god of war, Mars was often associated with death, bloodshed, and fire. What do you notice about it when you see it in the sky? Yes, it has a reddish-orange tinge, which might make an observer think of blood and war. As examined by the Viking craft, the Martian soil is composed of powdery rust (iron oxide). The air of Mars is primarily carbon dioxide (what you and I breathe out), the surface has no water, and the average surface temperature is about 75° F below zero. To top it off, Mars has no ozone layer, so living things would be seared by ultraviolet sunlight. With a world like this, life as we know it could not flourish. Even so, Mars has always sparked attention because of its unusual color—and the strange way it appears to move in the sky.

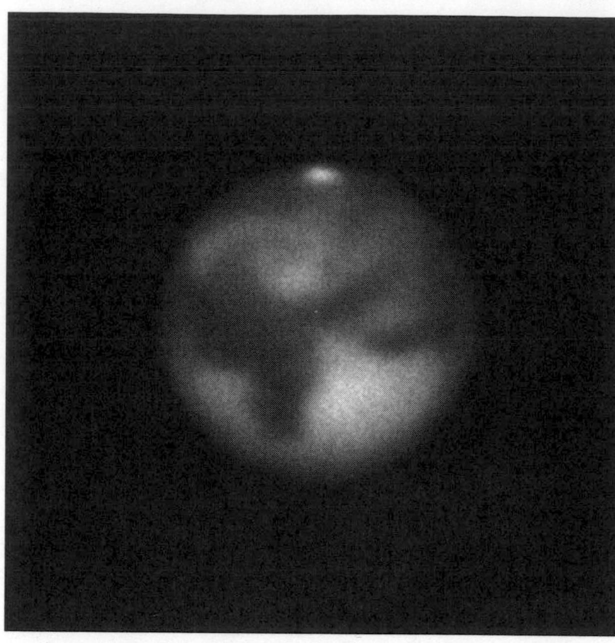

FIGURE 30-1. **Mars, the Red Planet. Note its variations in surface tone and the prominent polar cap. (Photo by Charles Layman)**

ACTIVITY 1

Watching the moon carefully from night to night reveals that it moves eastward about 12° a day against the stars. The planets do the same, but all at different rates. Because each planet is at a different distance from the sun, each moves at a different rate against the stars. Mercury, closest planet to the sun, orbits the sun in a speedy eighty-eight days. Mars, about one and a half times farther from the sun than Earth is, takes just under two years to orbit the sun. Saturn, much farther out still, orbits the sun in twenty-nine and a half years.

To get a better understanding of the planets' apparent motion against the backdrop of stars, go outdoors with a friend. Stand still as the friend walks around you to your left (counterclockwise) in a large circle. As your friend "orbits" around you, watch him or her against distant trees and bushes. Compared to the bushes, which way is your friend moving? You will see him or her moving *left* against the distant objects. In the sky, planets move the same way, only "left" is really "east." You can especially notice this if you follow a planet over many weeks.

But do planets always move eastward against the stars? Let's return to your circling friend and try something with a twist. This time, both you and your friend will move in a counterclockwise circle around a tree—you close to the tree and your friend far away. Make sure that your friend orbits the tree a little slower than you do. As you both orbit, watch your distant friend carefully against the background bushes and trees. What do you notice each time you catch up with the other person? It may be hard to notice at first, but as you approach and then pass your friend, he or she should appear to move momentarily to your **right** against the bushes. What you are seeing is an illusion, because your friend did not actually stop, move backward, and then start forward again. It is an illusion caused by your catching up with and overtaking your friend in your orbit.

In the sky, planets often show this same backward (**retrograde**) movement against the stars. Of all the planets, it is most easily seen with Mars when it is directly in line with Earth and sun (called an **opposition**), near the time it is closest to us (Figure 30-2). As Earth

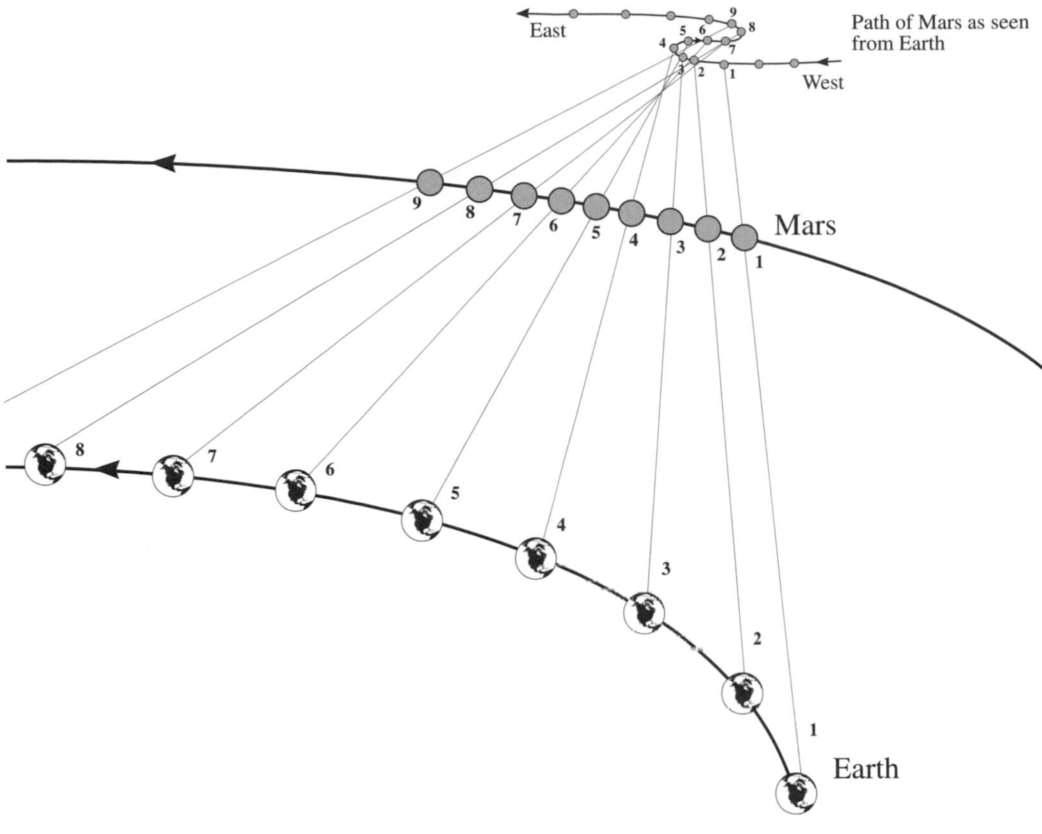

FIGURE 30-2. The retrograde loop of Mars is caused by Earth catching up with and passing Mars in its orbit. Sighting toward Mars when it is closest to Earth (at opposition) will reveal its westward loop (connect like numbers) as seen against the background stars (each planet is shown at one-week intervals).

approaches Mars, the usual eastward motion of the red planet appears to slow down and stop (when Mars is said to be stationary in the sky). Mars then appears to move **westward**—in retrograde (backward) motion—as Earth reaches its closest point to Mars and then passes the red planet. Mars will appear to travel west for several weeks. Finally, after Earth begins to recede from Mars, the red planet again appears stationary and then moves eastward again.

ACTIVITY 2

As Earth approaches and then passes Mars, the red planet forms a small loop among the stars, called a **retrograde loop** (Figure 30-3). At each Martian opposition, the shape of its retrograde loop is different, as it is related to the geometries of the orbital paths of both Mars and Earth. By carefully watching and plotting on a star chart the position of Mars during an opposition, you can make this retrograde loop yourself. To see how this works, use the charts in Figure 30-4 of a hypothetical Mars retrograde loop, to draw one of these unusual celestial shapes.

Beginning with chart 1, trace the stars onto tracing paper using a pencil. Overlay the traced chart 1 onto the stars of chart 2. What happened to one of the stars? This "star" is Mars, and it's not in the same spot as on chart 1. Continue overlaying the paper on charts 1 through 8, marking Mars each time with your red pencil. After the last chart, connect all the Mars points, and you will have a retrograde loop.

FIGURE 30-3a. A several-minute exposure of Mars's 1990–91 retrograde loop through Taurus (Pleiades at upper right of loop and Orion to lower left) taken using a GOTO planetarium projector.

FIGURE 30-3b. A similar Mars retrograde loop (from 1992–93), though of a different shape. Notice the stars of Gemini just above the loop.

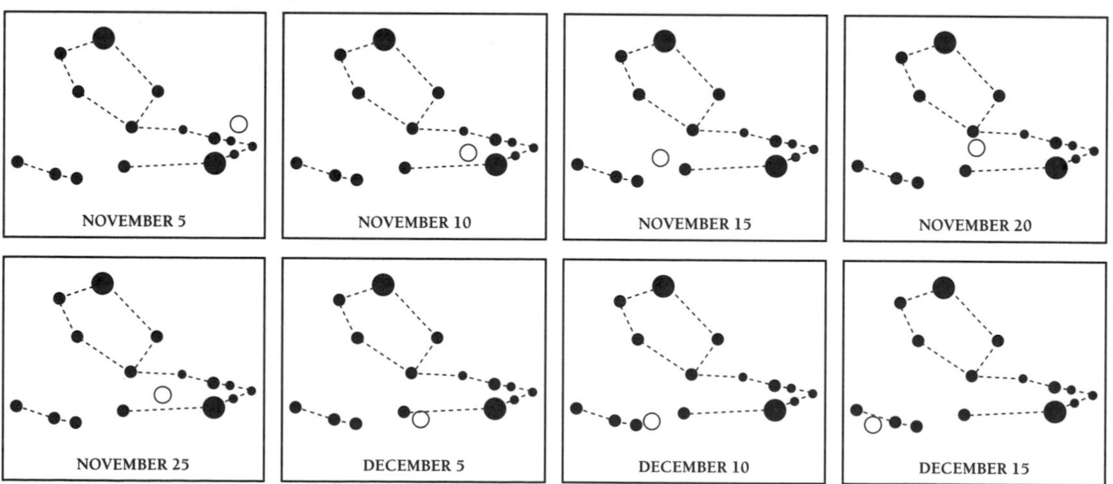

FIGURE 30-4.
With tracing paper, trace the star field in Chart 1. Then overlay the paper on Chart 2 and find the "star" that has moved. Continue to do so for all charts and connect the moving "star," Mars, in a retrograde loop.

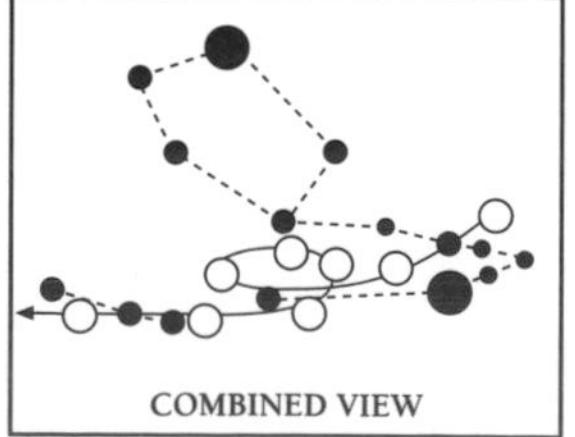

To find out when the next opposition of Mars will be, consult Table 30-1. Once you've followed the red planet's dance among the stars, you will understand why it caused such wonder among the ancients.

TABLE 30-1

This table provides approximate dates for retrograde loops of Mars, 1995–2007. The chart includes the Mars distance from Earth at opposition and its constellation positions during these times. The actual date of opposition is *during* the retrograde loop.

DATE	CONSTELLATION(S)	EARTH–MARS DISTANCE AT OPPOSITION (MILLIONS OF MILES)
Winter–Spring, 1995	Leo–Cancer–Leo	63
Winter–Spring, 1997	Virgo–Leo–Virgo	61
Winter–Spring, 1999	Libra–Virgo–Libra	54
Spring–Summer, 2001	Sagittarius–Ophiuchus–Sagittarius	42
Summer–Fall, 2003	Aquarius	35
Fall–Winter, 2005	Taurus–Aries–Taurus	44
Fall–Winter, 2007	Gemini–Taurus–Gemini	55

31 Jupiter

A Dynamic Planet

Jupiter has been visited by four unmanned U.S. spacecraft: Pioneers 10 and 11 and Voyagers 1 and 2. All sent back spectacular photographs of Jupiter. The United States has now sent the unmanned space probe GALILEO to Jupiter in an effort to answer more of the many questions we have about the planet.

Level: Intermediate and advanced

Objective: To observe the planet Jupiter

Materials:

- A telescope
- Photocopies of Jupiter
- Observation Form (included with this activity)
- One of the following periodicals:
  Sky & Telescope
  Astronomy
  Old Farmer's Almanac
  Observer's Handbook
  Sky Calendar
  Astronomical Calendar

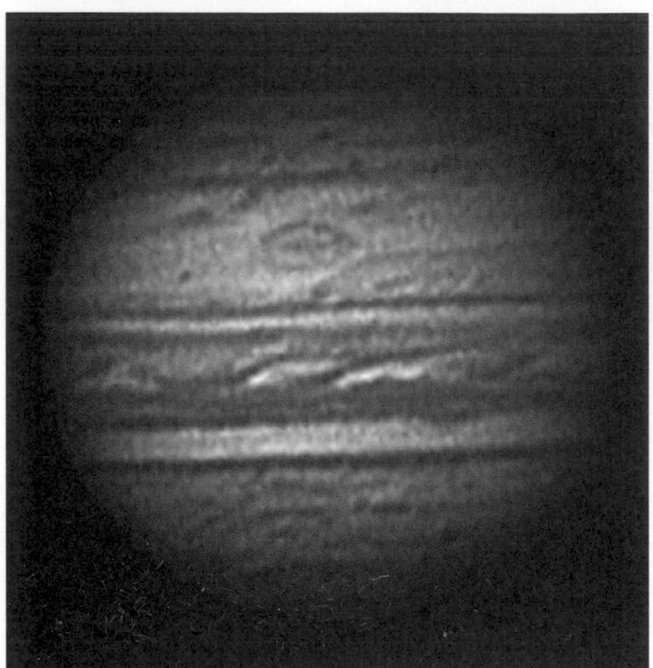

FIGURE 31-1 (above) Jupiter, the largest planet in the solar system. Note the Great Red Spot. (Photo by Gregory Terrance)

BACKGROUND

Largest of the nine planets orbiting the sun is mighty Jupiter. Jupiter measures 89,000 miles in diameter at its equator, more than eleven times the diameter of Earth.

Its huge diameter is only one way that Jupiter differs from our home. While Earth is encircled by a relatively thin atmosphere of mostly nitrogen and oxygen, Jupiter (Figure 31-1) is engulfed in an impenetrable atmosphere of hydrogen, helium, methane, and ammonia, all poisonous to life. Because of this dense atmosphere, astronomers have no way of knowing what lies beneath the clouds of Jupiter. Some scientists believe that the solid portion of Jupiter may be no larger than Earth, while others think the planet does not even have a rocky core.

Jupiter's atmosphere puts on a spectacular show through telescopes, revealing parallel dark belts and bright zones. Most prominent are two parallel dark belts crossing the middle of the planet. These are the north and south equatorial belts, identified in Figure 31-2.

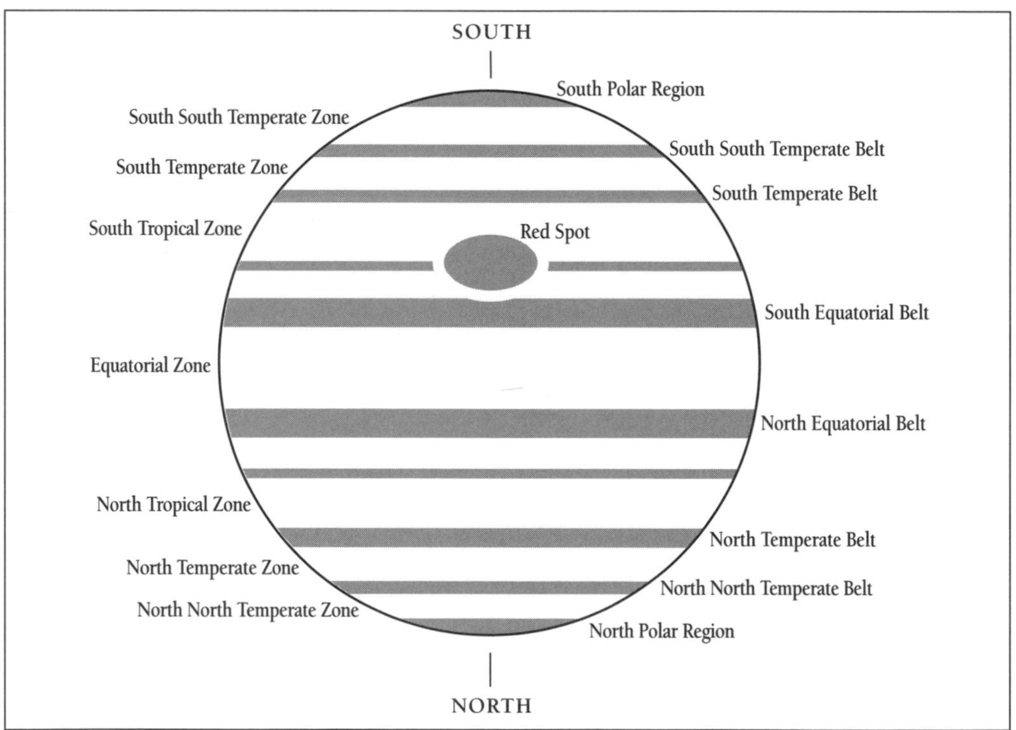

FIGURE 31-2 (right). Diagram showing the most prominent belts and zones in Jupiter's atmosphere. Compare this view with the photographs.

ACTIVITY

Is Jupiter visible in the sky tonight? To find out, check Appendix 4 at the end of this book or one of the publications in the materials list at the beginning of this activity. If so, and if the sky is clear, you can't miss it. Jupiter is the fourth brightest object in the sky, exceeded only by the sun, moon, and Venus.

Jupiter Observation

Date: ______________________ Time: ______________________

Telescope: ______________________ Power: ______________________

Notes: ______________________

____

____

____

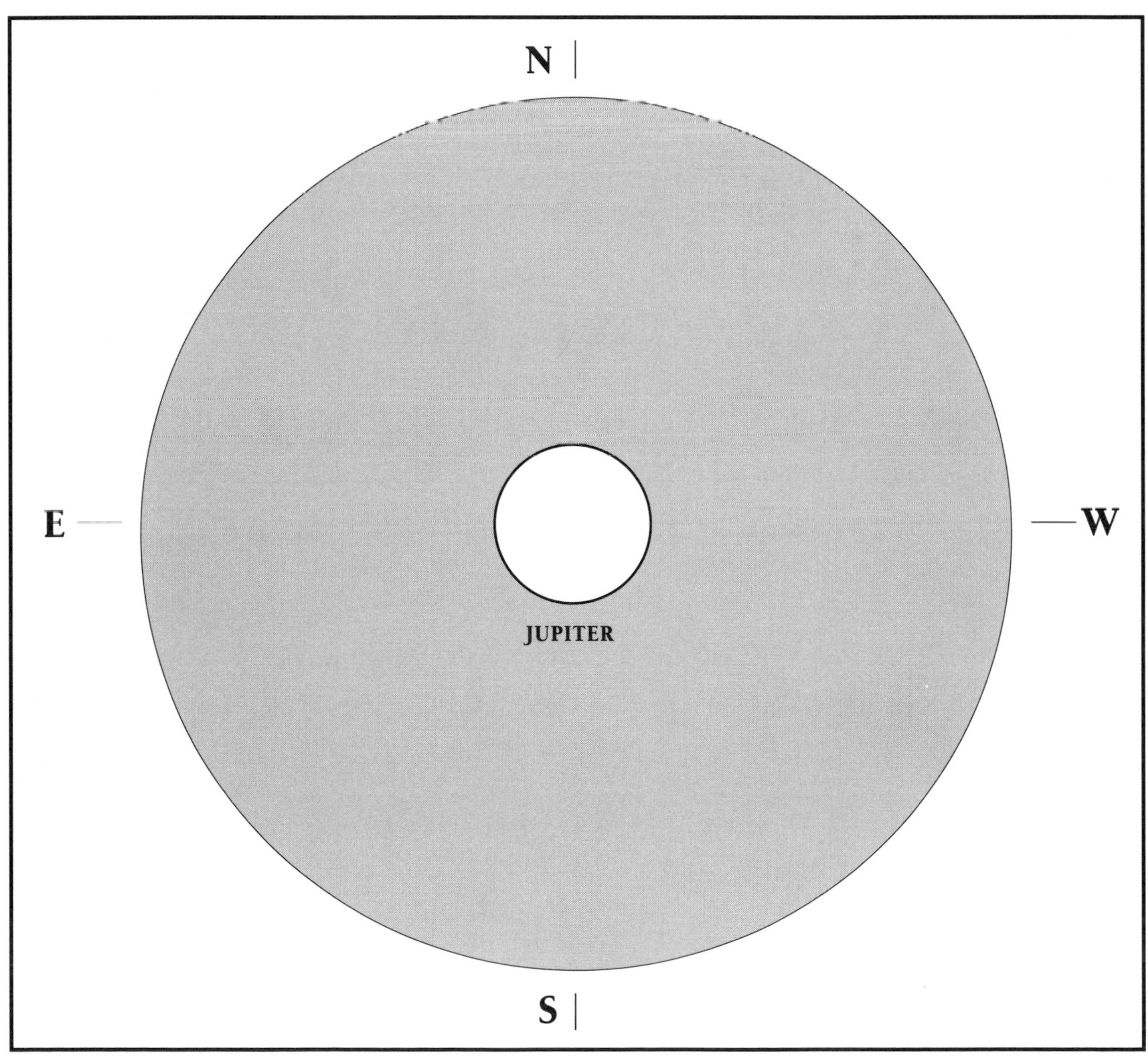

With an eyepiece in your telescope that produces between 50-power (50X) and 100-power (100X), aim toward Jupiter. Can you see the dark belts and bright zones?

Seeing the belts and zones of Jupiter is not as easy as it sounds. Depending on the atmospheric conditions of Earth, only one or two belts may be seen. And even then, they may be noticed only after five to fifteen minutes of concentrated viewing. To make matters more difficult, the entire planet might appear upside-down from Figure 31-2, depending on the type of telescope you have.

Using the Jupiter Observation Form, carefully draw all of the belts and zones that you can detect. Which appear most prominent? Which are the most difficult to see? Do you notice any color?

Even more difficult to see is Jupiter's famous Great Red Spot, a huge cyclonic storm that measures three times the size of Earth. Though it is visible with 100-power (100X) in telescopes as small as 3 inches across, the Red Spot is rarely red. Its tint is usually more of a pale pink or orange. To spy the Red Spot will require a slow and methodical search; a quick glance just causes it to blend into the surroundings. Of course, the Red Spot can't be seen at all if it happens to be on the far side of the planet at the time you are looking. If so, remember that Jupiter takes a little under ten hours to complete a rotation. If you can't see the Red Spot when you first look, come back out about five hours later and try again.

Mighty Jupiter serves up dramatic views through just about all telescopes, small and large. By studying it up close, you will be able to see just how dynamic the largest planet is.

32 Jupiter's Family of Satellites

Level: Intermediate and advanced

Objectives:
- To observe Jupiter's moons
- To determine which moon is which

Materials:
- A telescope
- Photocopies of Jupiter's Moons
- Observation Form (included with this activity)

BACKGROUND

In the last activity, you met Jupiter, largest of all planets, and its unique cloud system of dark belts and bright zones. No doubt as you viewed and drew Jupiter through the telescope, you noticed some star-like points on either side of the planet, like those seen in Figure 32-1. These are not stars at all but rather the brightest of Jupiter's family of no fewer than sixteen natural satellites, or moons.

Observation of Jupiter's Moons

Telescope: ______________________ Power: ______________________

Notes: ______________________

____

____

____

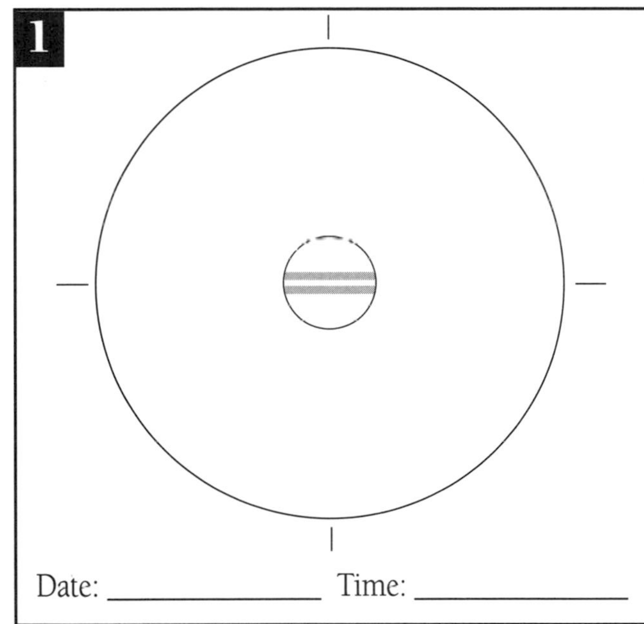

Date: __________ Time: __________

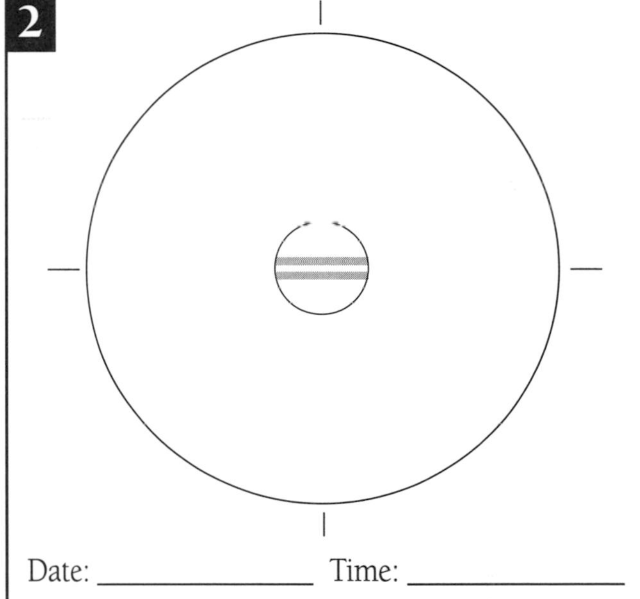

Date: __________ Time: __________

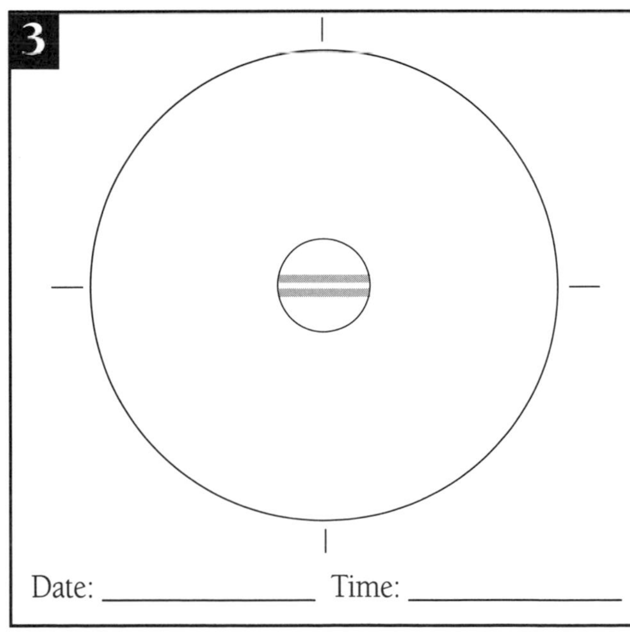

Date: __________ Time: __________

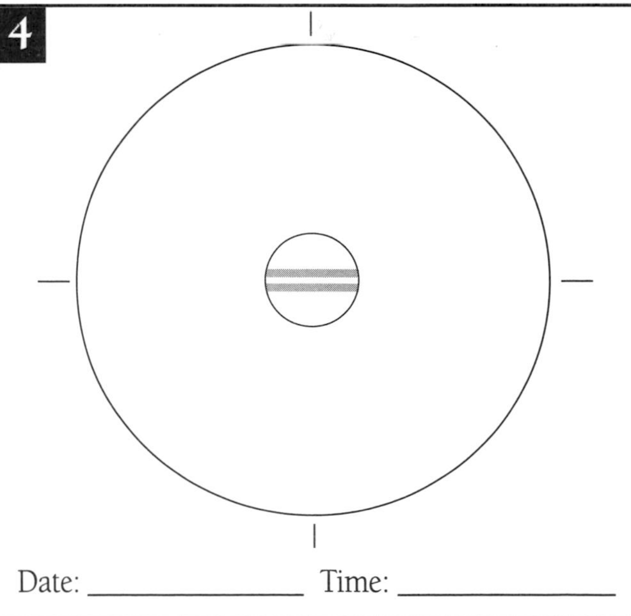

Date: __________ Time: __________

Up to four of Jupiter's satellites are visible through small telescopes and binoculars. All four were discovered by Galileo in the early seventeenth century when he turned his first crude telescope skyward. The moons were named **Io**, **Europa**, **Ganymede**, and **Callisto** after mythological acquaintances of the Roman god Jupiter. Collectively they are referred to as the **Galilean satellites**.

ACTIVITY

Even a casual glance at Jupiter will reveal the four Galilean satellites as small "stars" extending to either side of Jupiter and parallel to the planet's belts and zones. Sometimes you will see two on one side and two on the other; at other times, there may be three on one side and one on the other. Perhaps only three, or even only two, will be visible, the other one or two hidden from view either behind or in front of the planet.

Which moon is which? With just a quick look, it is just impossible to tell. But by tracking the satellites for a couple of hours each night for several days, it should be possible to tell which is which. No two of the Galilean satellites revolve about Jupiter at the same rate. The closest of the four to Jupiter, Io, takes only one day, eighteen hours to complete an orbit. Next, in order out from the planet, is Europa at three days, thirteen hours; Ganymede at seven days, four hours; and Callisto at sixteen days, seventeen hours.

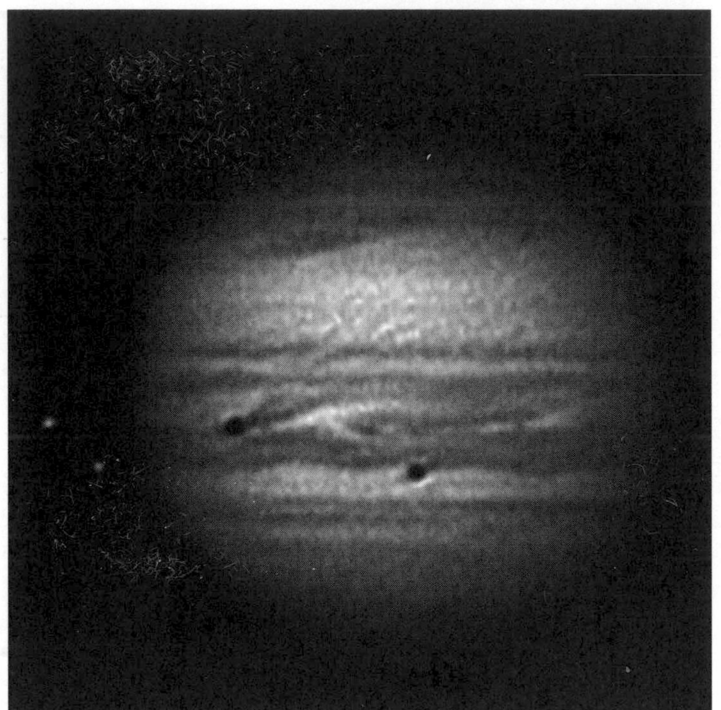

FIGURE 32-1 The giant planet Jupiter attended, in the foreground, by two of its Galilean moons. (Photo by Gregory Terrance)

Each clear night over a two-week period, observe the satellites closely using a telescope at 50-power to no more than 100-power. Accurately draw their positions on the Jupiter's Moon Observation Form, noting the exact time and date of the observation. Make four observations each night, each separated by at least an hour.

When the record is complete, compare the results with the order outlined above. Can you tell which satellite was which? It may still prove difficult. If so, consult either *Sky & Telescope* magazine or *Astronomy* magazine (probably available at your local library). Each monthly issue of these magazines includes charts showing the nightly positions of Jupiter's moons, making it easy to determine one from the other.

By comparing your drawings with the magazine charts, estimate how long it takes each of the Galilean satellites to complete its orbit. (Make certain the charts are oriented the same as your drawings.) Compare your results with the times listed earlier. How close are your estimates?

Watching the dance of the moons around Jupiter is a fascinating way of seeing that the universe is always changing, always moving—an active universe indeed.

33 Ring around the Planet

Observing Saturn

Level: All

Objective: To observe Saturn and its ring system

Materials:

- A telescope
- Photocopies of Saturn
- Observation Form (included with this activity)
- One of the following periodicals:
  Sky & Telescope
  Astronomy
  Old Farmer's Almanac
  Observer's Handbook
  Sky Calendar
  Astronomical Calendar

FIGURE 33-1 Saturn, the solar system's ringed wonder. (Photo by Gregory Terrance)

BACKGROUND

Lying beyond the orbit of Jupiter at nearly twice the distance from the sun, Saturn is the farthest member of the solar system that is visible to the naked eye. Before the invention of the telescope, Saturn was seen as the slowest of the planets (or "wandering stars," as they were called). By watching it slowly change its position against the background stars, astronomers learned that Saturn takes $29\frac{1}{2}$ years to go once around the sun.

With the invention of the telescope, Saturn, like the other planets, came under close scrutiny. In 1610, Galileo became the first person to see Saturn through a telescope, but the view was not impressive. He noted that there was something peculiar about this distant world, but he couldn't quite pinpoint it. In his observation record book, Galileo wrote that Saturn appeared to have "ears," and he accompanied the description with a drawing that looked more like Mickey Mouse than a planet.

Saturn remained a mystery until 1656, when Dutch astronomer Christian Huygens first confirmed that Saturn did not have ears but rather a ring around it. At first glance Saturn appears encircled by a single wreathlike ring. A close-up view, as in Figure 33-1, shows that Saturn's ring is actually divided into several sections. Figure 33-2 shows the ring's major segments. When the Voyager I and Voyager II spacecraft flew by the planet in the 1980s, they returned pictures of the rings that revealed them to be divided into thousands of "ringlets."

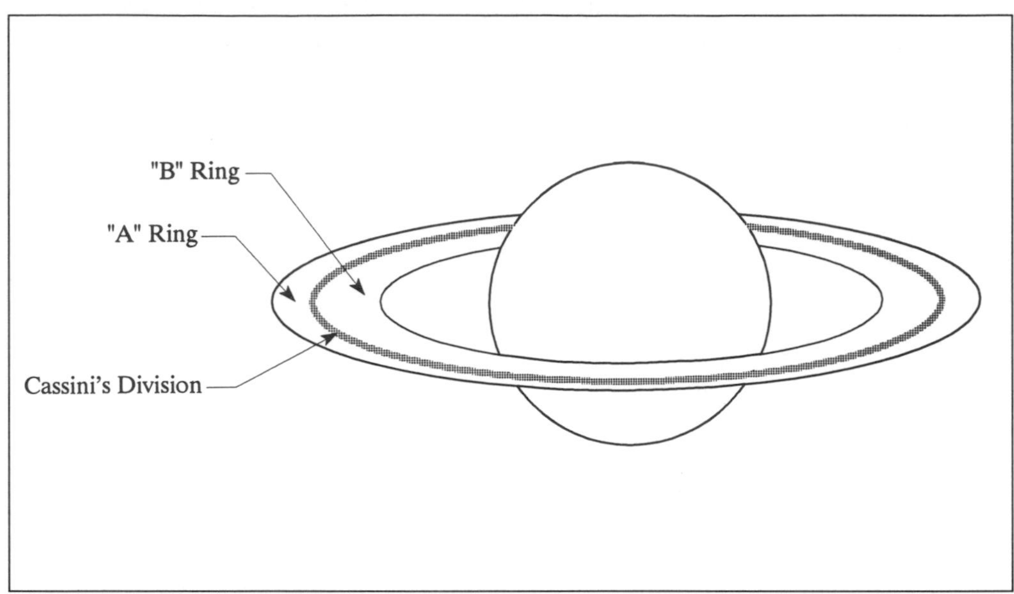

FIGURE 33-2. **Diagram showing Saturn's prominent A and B rings.**

ACTIVITY

Many people new to astronomy are under the mistaken impression that you need a big telescope to see Saturn's famous system of rings. Nothing could be further from the truth. Just about any telescope capable of magnifying 25 times (25X or 25-power) can resolve the rings of Saturn.

First consult Appendix 4 or one of the publications in the materials list at the beginning of this activity. Is Saturn going to be visible tonight? Until after the turn of the century, Saturn will be traveling through the dim zodiacal constellations of autumn. Locate its current constellation on the appropriate star map found earlier in this book, go outdoors, and have a look.

Once you spot Saturn, what should you expect to see? With the naked eye or binoculars, Saturn appears as a slightly beige or yellowish "star." Because the constellations it passes through between now and the year 2000 contain no bright stars, Saturn's presence will be easy to detect.

With a telescope and an eyepiece magnifying at least 25-power (25X), Saturn's true identity comes through. Can you see the rings? What do they look like? What color are they? Can you see the dark line cutting the rings in half? This line (actually, it's a gap) is called **Cassini's Division** after the astronomer who first discovered it. As shown in Figure 33-2, Cassini's Division separates the so-called "A" ring from the "B" ring. Take a look at the planet itself. Can you make out any markings in Saturn's atmosphere? Record your observations on a copy of the Saturn Observation Form. Draw what you see, and then compare your view with that of Galileo.

If you do this activity in 1995 or 1996, you're in for a big surprise. During those years, when you look at Saturn, you will see a ring-less planet. Don't worry. Twice in Saturn's orbit of the sun, we see Saturn's rings **edge-on** for a period of time. Although they are thousands of miles wide, Saturn's rings are believed to be only a couple of thousand **feet** thick. As a result, when they appear edge-on, the rings disappear from our view. By late 1996, Earth and Saturn will have progressed far enough along in their orbits to cancel out the effect, and we will have our ringed wonder back again.

Saturn Observation

Date: ______________________ Time: ______________________

Telescope: ______________________ Power: ______________________

Notes: ____

____

____

____

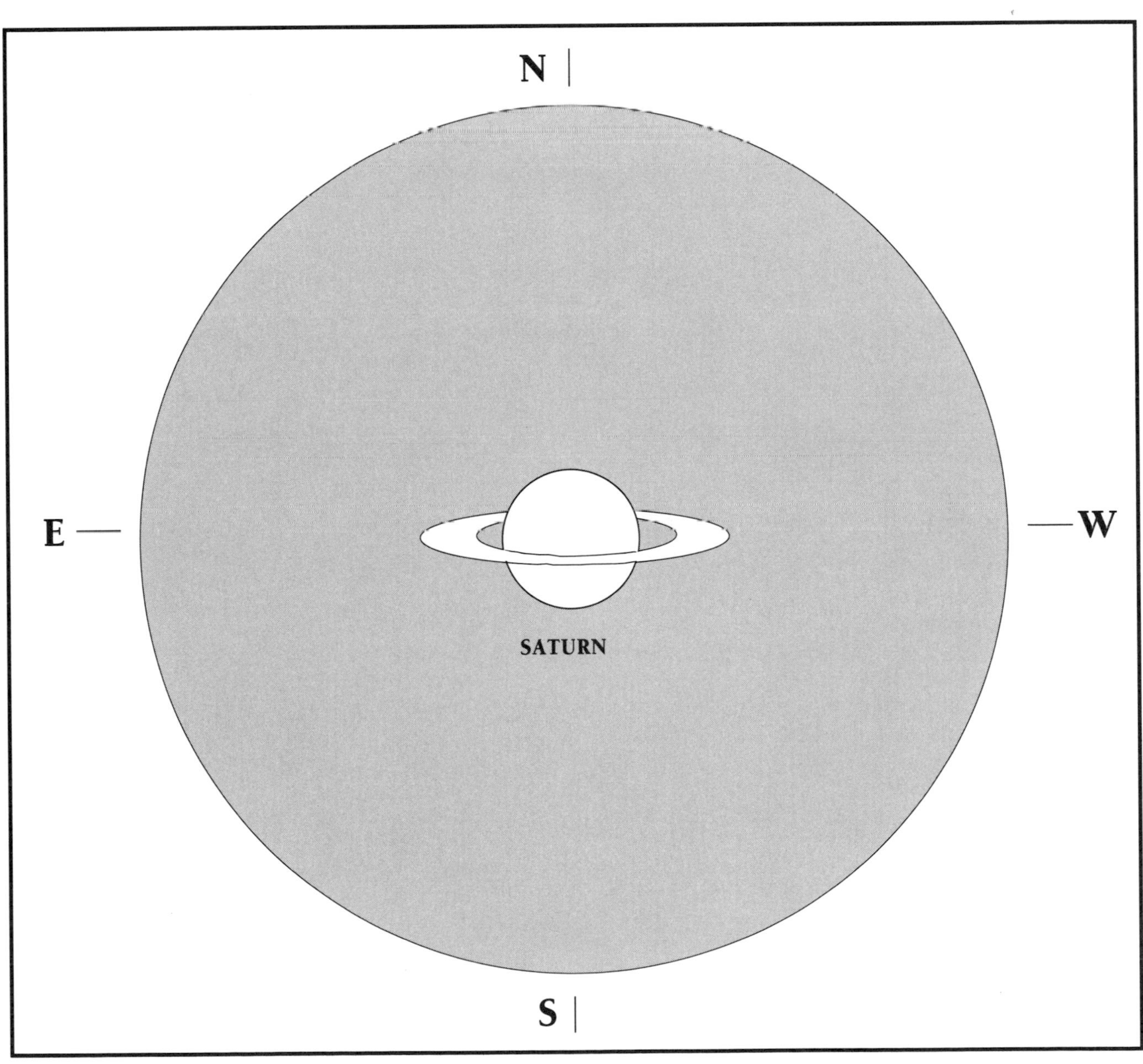

34 Weighing Yourself on Other Planets

Level: All

Objective: To understand the differences in gravity on each planet

Materials: A pen or pencil

BACKGROUND

Imagine picking up a tennis ball and dropping it from a third-floor window. Watching carefully, someone on the ground may notice that the ball gradually speeds up as it falls. On Earth, each second the ball falls (up to a certain point), its speed increases by an additional 32 feet per second every second. Here we are witnessing an effect of Earth on the ball. Similarly, each time we stand on a bathroom scale and check the reading, we see an effect of Earth on us. Because Earth has a certain size and contains a certain amount of material (called its **mass**), it attracts other objects, such as the moon, animals, falling objects, and people. In the mid to late 1600s, physicist Isaac Newton studied the attraction of Earth for the moon and reasoned that **every object** in the solar system attracts **every other object**. This "attraction," often thought of as a "pull," is called **gravitational force**, or simply **gravity**. Gravity gives things their weight and is responsible for causing objects to fall.

Deimos, a moon of Mars that is only about 8 miles across, has a gravitational field that is more than 2,900 times weaker than Earth's. A person who weighs 150 pounds on Earth would weigh about one ounce on Deimos. If you were on Deimos, you could propel yourself into orbit simply by reaching a running speed of about 7 miles per hour.

As Newton realized, the moon "falls" (is attracted) toward Earth in the same way a plummeting ball or apple does. How long it takes an object to fall a certain height depends on a planet's gravity. For example, on Earth, a tennis ball dropped from a height of three stories takes about 1.4 seconds to reach the ground. But on the moon, it would take 3.5 seconds, more than twice as long. On Jupiter, the drop would take less than a second. The differences in time are due to the fact these celestial bodies have different sizes and masses and thus possess different amounts of gravity. A typical major league home-run ball on Earth would travel nearly 1 mile if hit on the moon, due both to the fact that the moon is airless and that it has only one-sixth the gravity of Earth.

ACTIVITY

In this activity you will be whisked through the solar system, stopping at the sun, moon, and planets to determine your weight on each of these worlds. You may ask, "Will my size change if I lose or gain weight by visiting another planet?" At first it might make sense that you would, but in reality you won't. By going to another planet, you are changing the object that causes you to have a certain weight (the planet itself) and are not "dieting." The amount of matter that composes you (your *mass*) remains the same.

Table 34-1 lists the sun, moon, and planets next to a column of numbers. Notice that all of these numbers are different (with the coincidental exception of Mercury and Mars). Each number represents how much more or less that planet's gravity is in comparison with that of Earth. For example, imagine standing on Venus instead of Earth. As the table shows, the force with which Venus pulls on you (your weight) will be only 0.91 that of what you would

TABLE 34-1

Object ____________________

Weight on Earth ____________________

CELESTIAL BODY	RELATIVE GRAVITATION FACTOR	NEW WEIGHT
Sun	27.9	
Mercury	0.38	
Venus	0.91	
Moon	0.17	
Mars	0.38	
Jupiter	2.54	
Saturn	1.00	
Uranus	0.91	
Neptune	1.19	
Pluto	0.06	

normally feel here on Earth. You would weigh less. But if you went to Jupiter, your body would instead weigh 2.54 times what it does on Earth. (These numbers also show how much more or less a falling object will speed up, compared with the same falling object on Earth, and thus represent the **acceleration due to gravity**.)

Write your weight in the appropriate box in Table 34-1. To find your weight on the sun, moon, and planets, multiply your Earth weight by the number following each. Fill in the blank spaces under "your new weight" as you go down the column. Where in the solar system will your weight be its lowest? Where will it be highest? Notice that your weight on Pluto would be less than it is on the moon.

35 Looking for Meteors, Meteor Showers, and Fireballs

Level: All

Objective: To observe meteor showers and other meteor phenomena

Materials: A clear, dark, moonless night sky, far from city lights, during a major meteor shower

FIGURE 35-1. A bright meteor leaves its characteristic streak among the stars of the Little Dipper (the North Star is in the very lower right corner). (Photo by Dennis Milon, August 12–13, 1980)

FIGURE 35-2. Leonid meteors darting through the sky, as shown here in an 1833 painting of unknown origin. The Leonid shower can number in the tens of thousands of meteors per hour every 33 1/4 years. The next anticipated burst of this shower is to be in November 1999. (NASA photo)

BACKGROUND

As you learned in Activities 4 through 7, people long ago imagined the stars and constellations to be characters in their favorite stories. Occasionally, as they gazed upward, they would spot rapid streaks of light darting across the sky, lasting but a second or two (see Figure 35-1). At other times, the night sky would be filled with these streaking lights (see Figure 35-2). Many ancients thought this meant that their friends were falling out of the sky or, worse yet, that the whole sky was falling down to Earth.

Incorrectly called "shooting stars," these brief dashes of light are actually small rocky debris passing through Earth's atmosphere some 50 to 70 miles above the ground. They are called **meteors**. Ranging in size from a pea to a grain of sand, these rocks travel at speeds ranging from 30 to 50 *thousand* miles per hour! Moving so fast through the air, meteors experience air friction, which causes them to heat up, glow, and either shatter or disintegrate entirely. (To understand a little about friction, rub your hands together very quickly; what happens?)

Simultaneously, the rapidly moving stone also heats up air molecules, causing them to glow as well. As a result, what you see is a quick streak of light in the night sky as the meteor passes through our atmosphere. The smallest of these meteors, called micrometeorites (see Activity 36), drift slowly toward Earth, whereas meteors about the size of your fist and larger may hit Earth after a colorful passage through the air. Sometimes called "nature's fireworks," such meteors are most commonly known as **fireballs**. Should the meteor "pop" and crack into visibly colorful pieces during its atmospheric entry, it is called a **bolide**.

During its history, Earth has experienced a barrage of meteor impacts that have left countless pockmarks, called CRATERS. After 4.5 billion years of resurfacing, weathering, and erosion, many craters have filled in, making them virtually undetectable. Some, though, remain nearly pristine, such as Meteor Crater near Flagstaff, Arizona (Figure 35-4). Nearly 600 feet deep and 1 mile wide, this impressive sight was created about 49,000 years ago by a meteor weighing several thousand tons.

The major meteor showers, shown in Table 35-1, surprisingly come primarily from comets. Upon nearing the sun, a comet, also known as a dirty snowball, is heated and slowly vaporizes, creating a gaseous tail and a stream of small chunks of rock (see Activity 37, especially Figure 37-3). This region of cometary debris then remains in its own orbit; individual particles are called **meteoroids**. As Earth passes through this stream of debris, the particles, called **meteors**, enter Earth's atmosphere. The ones that do not disintegrate but survive to impact Earth are called **meteorites** (see Figure 35-3). Meteor showers and sporadic meteors (about 7 per hour per night are visible) together contribute some 70,000 tons of disintegrated rock to Earth's atmosphere each year! Meteorites are generally indistinguishable from Earth's rocks, although most Earth stones, unlike meteorites, show signs of weathering and erosion, often giving them rounded shapes. Thus, it often requires careful scientific studies to reveal if a rock is meteoric or not.

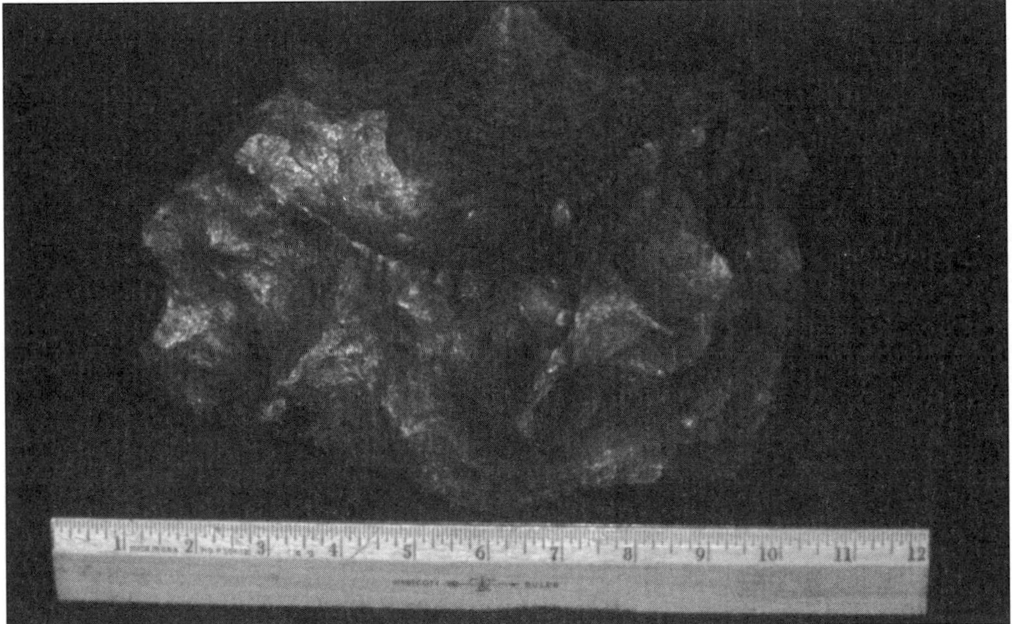

FIGURE 35-3. A typical nickel-iron meteorite. This one, weighing some sixty pounds, fell in Texas about 10,000 years ago. It is on display at New York's Vanderbilt Planetarium.

Though most meteors in these showers usually burn up in our atmosphere, occasionally a meteor will impact Earth's surface. Such larger meteors usually don't originate from any particular comet, but may instead be remnants of asteroid collisions in the Asteroid Belt (between Mars and Jupiter). In fact, some meteorites found on Earth are believed to be ejected moon rocks kicked up during a meteor impact on the moon, some 240,000 miles distant. Some are even from Mars, some 40 million miles from Earth *at its closest.*

The main classes of meteorites—**stones**, **stony-irons**, and **irons**—are determined by the constituents of the meteorite. Stony meteorites, the most abundant seen to fall, are similar to Earth rocks and therefore difficult to find. Some contain a great deal of carbon, an element that's present in living things. In fact, some contain amino acids, the building blocks of life. Having remained unchanged since their formation 4.5 billion years ago, stony meteorites are the oldest material ever to be held by humans. Irons, on the other hand, are very dense mixtures of metals, mostly iron and nickel, that have experienced significant heating and melting since their formation. Stony-irons are a mix of the two. Iron-rich meteorites attract compass needles or cause watch windings to spin.

An odd meteor impact in 1992 crumpled the trunk of a parked car (see Figure 35-5) belonging to Michelle Knapp of Peekskill, New York. The damage was initially considered to be the work of a vandal, but the perpetrator was soon discovered to be a twenty-seven-pound

FIGURE 35-4. A crater still well preserved today near Flagstaff, Arizona, is the legacy of a multi-megaton bomblike meteor impact some 49,000 years ago. (Photo courtesy of Valda S. Eyrich, Meteor Crater Enterprises)

FIGURE 35-5. A car belonging to Michelle Knapp of Peekskill, New York, was the victim of meteor impact in October 1992. (Photo courtesy of John E. Bortle, W.R. Brooks Observatory)

stony meteorite whose path was witnessed by many along the eastern seaboard on the evening of October 9. Ironically, Ms. Knapp quickly received handsome offers for both the meteorite and the car from scientists, museums, collectors, and even London's Sotheby's auctioneers!

A rare motion picture made in August 1972 in the Grand Teton mountains of Wyoming captured a meteor as it passed through our atmosphere some 36 miles above the ground never striking Earth's surface! This incredible footage, taken by James Baker of Omaha, Nebraska, was invaluable to astronomers, who used it to track the meteor's path, size, and altitude. The 13-foot-long meteor left a clear smoke trail visible for nearly thirty minutes. The meteor itself, visible during daylight for no more than a minute, caused several sonic booms over Utah, Wyoming, and even Alberta, Canada, before leaving Earth's atmosphere.

ACTIVITY

Venture outdoors on a clear, moonless evening as far from city lights as possible. Pick a night when you expect meteor showers. Table 35-1 lists major annual meteor showers, their respective parent comets, and the anticipated number of meteor streaks per hour. You may see longer and brighter meteors prior to midnight, with dimmer but more frequent meteors visible well after midnight, when Earth is rotating into the main meteor stream.

Take careful notes on the time, direction, brightness, and colors of the meteors you see. Look for the **radiant**, the constellation from which all the meteors of a particular shower appear to emanate. This will give you a general idea of the location in space from which a particular meteor swarm will occur each year. The radiant also gives the name of the meteor shower. For example, the constellation of Orion gives its name to the Orionid meteor shower. To best determine the shower's hourly rate, observe with many people, each facing a different direction. Patience is important, so plan to observe for at least an hour and dress appropriately.

Speaking of hourly rates, one of the most amazing showers is the Leonid meteor shower, which has been known to flare up from a low 10 meteors per hour to an incredible 10,000 to 100,000 meteors per hour once every thirty-three years. The next anticipated Leonid burst of this magnitude is expected in mid-November of 1999. Don't miss it!

TABLE 35-1

Principal Shower	Approximate Dates (Normal Limits)	(Maxium)	Approximate Hourly Rate of Metrors	Known Parent Comet
Quadrantids*	Jan. 1–6	Jan. 3	60	—
Lyrids	Apr. 19–25	Apr. 22	10	Thatcher 1861 I
Eta Aquarids	May 1–10	May 6	35	Halley (periodic)
Delta Aquarids	July 15–Aug. 15	July 29	20	—
		Aug. 2	10	—
Perseids	July 23–Aug. 20	Aug. 12	75	Swift-Tuttle (periodic)
Orionids	Oct. 16–27	Oct. 22	25	Halley (periodic)
Taurids	Oct. 20–Nov. 30	Nov. 5	10	Encke (periodic)
Leonids	Nov. 15–20	Nov. 17	10	Tempel-Tuttle (periodic)
Geminids	Dec. 7–15	Dec. 13	85	Asteroid #3200, Phaethon

*Named after the former constellation Quadrans Muralis, now visible in Boötes.

36 Looking for Micrometeorites

Level: All

Objective: To collect and study microscopic debris from space

Materials:

- An aluminum pan, about 12 inches by 18 inches
- A small bowl
- A magnet
- A brick
- Two clear plastic bags
- One quart of distilled water
- Aluminum foil
- A microscope or hand-held magnifying glass (optional)

BACKGROUND

Micrometeorites are very tiny meteorites. They can be either the debris of comets passing near the sun and Earth or leftover dust from the formation of the solar system. When Earth passes through the debris trail left by a comet, skywatchers are treated to meteor showers composed of small rocks—ranging in size from as big as your pinky nail to as small as a grain of sand—that flare up as they disintegrate in Earth's atmosphere. Some of these meteor bits, though, are smaller; the size of the dot on the letter "i" and even tinier. In this activity, you will learn how to capture some of these minuscule outer-space particles (which are called micrometeors while they are airborne and micrometeorites after they touch down on Earth).

Micrometeors, unlike larger meteors, do not burn up in the atmosphere as they approach Earth because they are too lightweight to do so. In the same way that a piece of plastic foam floats in a pool of water, micrometeors float upon our atmosphere. They have such little weight that the air can support them.

The greatest concentration of micrometeors occurs around the appearance of a major meteor shower (see Table 35-1 in the previous activity for a list of shower dates). Thereafter, the particles can remain suspended in the air for weeks or months until they are washed toward Earth's surface by wind, rain, or snow. It is estimated that some 100 tons (about 200,000 pounds) of micrometeors enter Earth's atmosphere each day, with more entering during large meteor showers. Their presence is not normally noticed because they leave no visible trail.

ACTIVITY

The easiest way for micrometeors to reach Earth's surface is to be carried down by a rainstorm. Thus your best opportunity to collect these extraterrestrial grains will be during a rainy period. You should have even better success if you try to collect the micrometeorites during a rainy period that coincides with a major meteor shower.

Start by lining the bottom of your 12-inch by 18-inch aluminum pan with aluminum foil. To serve as an isolated weight, place the brick in a plastic bag, tie the bag closed tightly, and place it in the pan. Now put the pan in as high and clear an outdoor location as possible (the best spot is on a rooftop in a clearing). Allow the pan to collect rainwater before, during, and after a meteor shower.

Several days after the peak of meteor activity, it's time to see if you've captured anything. Put the magnet in a clear plastic bag. Run the bagged magnet through the rain-filled pan to attract metallic micrometeorites. The bag makes it easier to remove the particles from the magnet. Line the small bowl with aluminum foil, fill it with distilled water, and put the bagged magnet and adhering particles into the bowl.

Carefully remove the magnet from its bag, allowing the particles to settle to the bowl's bottom. You can then collect the particles by letting the bowl stand covered with gauze for a few days and allowing the water to evaporate. (You can also collect the particles by pouring the water from the bowl into a foil-lined pot and then boiling off the water, but this is a less desirable method.) Similarly, evaporate or boil off the water from the large aluminum pan to collect the nonmetallic micrometeorites. Allow the particles to dry, being sure to keep them from dust.

Once you have collected and dried the particles, sort them by size. A magnifying glass or a microscope with a 100X eyepiece may be helpful in looking at the very small pieces. Though many of the particles you have collected will be from Earth (such as pollutants and dust), many will have originated from comets or interstellar space. If you collect over enough days, you should be able to notice that the amount of Earth debris is relatively constant, whereas the amount of space debris varies over the course of time.

The particles you collect may be from any of the three main classes of meteor material: stones, irons, and stony-irons. The metallic micrometeorites will most likely fall into the iron or stony-iron categories. Because of their magnetic properties and high iron-nickel content, these rocks are fairly easy to distinguish from terrestrial stones. The nonmetallic particles in your collection will most likely be of the stony classification. These are harder to distinguish from Earth stones. Geologists and planetary astronomers use sophisticated equipment to analyze the stones to determine their actual composition.

37 Dirty Snowballs?

Observing Comets

Level: All

Objective: To learn about and observe comets

Materials:

- A comet in the sky
- A pair of wide-field binoculars
- A telescope (optional)
- Seasonal star charts (see Activities 4 through 7)
- Activities 9, 10, and 13 (as references)

BACKGROUND

"On both sides of the rays of this comet were seen a great number of axes, knives, and blood-colored swords among which were a great number of hideous human faces with beards and bristling hair . . ." So says a sixteenth-century description of one of history's grandest celestial visitors, Halley's comet (Figure 37-1). The first documented sighting of Halley's comet was made in 240 B.C. The appearances of this and other comets were often considered omens of disaster. As recently as 1910, during that return of Halley's comet, fear caused many people to purchase "comet pills" to protect them from these allegedly harmful visitors. In this activity, you will learn the truth about such objects and how to enjoy their visits to your celestial backyard.

FIGURE 37-1. Comets were often perceived as harbingers of doom, as depicted here following a sixteenth-century description of Halley's Comet. (Courtesy of Vanderbilt Planetarium)

Bright comets are rare (the last was Comet West in 1976). Most comets are very faint, and you need optical assistance to easily view them. Obtain a list of current comets in the sky, or a detailed track of a bright comet against the stars, by consulting the ASTRONOMICAL CALENDAR, the SKY CALENDAR, SKY & TELESCOPE magazine, or ASTRONOMY magazine (see Appendix 1). These publications distribute comet news worldwide as backyard astronomers discover more and more of these exciting objects. For up-to-the-minute astronomical news reports, call SKY & TELESCOPE's skyline at (617) 497-4168.

The word "comet" derives from the Greek words for "star" and "hair." A comet—with its bright head and shining spread-out tail—was called a "hairy star" (Figure 37-2). A comet looks like this because of the sun's effect on it. When a comet is vastly far from the sun and beyond the view of Earth's largest telescopes, it is nothing more than a ball-shaped object composed of ice, rock, and dust—often called a "dirty snowball." This comet model, proposed by astronomer Fred Whipple, suggests that this ice includes water ice (like ice cubes in your freezer), carbon dioxide ice ("dry ice"), and ices of ammonia and methane. Intermixed are chunks of the solar system's oldest rock and dust. All of this material together forms the comet's **nucleus**—invisible from Earth because it is often a mere 5 to 30 miles across.

As the comet nucleus approaches the sun, solar heat and light begin to sublimate the ices of the nucleus (like the gas that "steams" off a block of dry ice), forming a type of "cloud" engulfing the nucleus. Called the **coma**, this cloud is typically about 65,000 miles across and, through a backyard telescope, often appears as a fuzzy star. The nucleus and coma form the comet's head.

The most spectacular part of the comet is its tail, composed of sublimate gases and escaping dust particles from the nucleus. Comets generally show two tails, one curved **dust tail** and a second (often bluish in color) straight **gas** or **ion tail**. Dust tails primarily indicate the path of the comet, whereas the gas tail indicates the radial path of the sun's rays (Figure 37-3). A comet's tail, always pointing away from the sun, can extend up to 100 million miles into space (as did that of comet Ikeya-Seki in 1965; Figure 37-4) and cover a good portion of the night sky. The tail of the Great Comet of 1843 stretched halfway across the sky (some 90°) and was an estimated 500 million miles long—long enough to stretch from the sun to Jupiter. It is no wonder that the ancients considered these beautiful objects harbingers of doom.

FIGURE 37-3. (Above) Comet Bennett 1970, photographed from Harvard, Massachusetts, by Dennis Milon, clearly shows a straight gas tail (the faint one) and a larger, brighter curved dust tail.

FIGURE 37-2. (Left) One of the more impressive comets in recent history was Donati's Comet, shown here. This lithograph, from the frontispiece of the 1877 edition of Amedee Guillemin's *Le Ciel*, shows the comet over Ile de la Cité in Paris on October 4, 1858, near the star Arcturus.

FIGURE 37-4. Comet Ikeya-Seki, as seen from the Catalina Mountains in Arizona on October 31, 1965. A rare "sungrazing" comet, Ikeya-Seki had an estimated tail length of over 100 million miles. (Courtesy of Dennis Milon)

Comet Observation

Comet's name: ______________________________

Date: ______________________ Time: ______________

Telescope: ______________________ Power: ______________

Constellation: ______________________ Comet's angular size: ______________

Was the comet visible to the naked eye? ______________________________

What was the comet's approximate magnitude? ______________________________

Did the comet show a tail? ______________ How long was it? ______________

Notes: ______________________________

____

____

____

ACTIVITY

Studying comets is not a difficult task using binoculars or a telescope. A good pair of wide-field binoculars should yield a beautiful view under a dark sky. The true difficulty with comets is finding them in the first place and finding a bright one. At any one time, there may be half a dozen comets in the sky, though most are too faint to be visible with anything less than large telescopes. Comet hunting takes hours of painstaking work and must be done under the best sky conditions possible. This is usually just after sunset or just before sunrise, because this is when a comet is close to the sun and would thus have a good-size tail. By the way, the first person to spot a new comet is awarded the ultimate prize: the comet bears the discoverer's name.

If a comet within the reach of binoculars or a small telescope is in the sky, jump at the chance to look at it. Document your observations on a copy of the Comet Observation Form. Try to determine its brightness (magnitude); consult Activities 9 and 10. If it is just barely visible to your naked eye from a dark rural location, you may estimate the comet at about magnitude 5.5. A comet as visible as a planet is much brighter, at about magnitude 2.0 to 4.0. The magnitude of a comet with a tail stretching across the sky may be much harder to estimate. Such a beautiful sight is a once in a lifetime event!

Using binoculars or a telescope, try to spot the bright coma. How wide is it? A good way of estimating this is by knowing the size of your binocular's field of view and then counting the number of comas that can fit into it. Similarly, try to measure the length of the comet's tail, either with optical aid or by using your hand and naked eye (see Activity 13). In this way, you can determine the angular size of the coma and tail of the comet. As you do, try to observe changes in the coma's brightness and tail length. Also note the path of the comet against background stars (see star charts in Activities 4 through 7), and perhaps try to plot it on a star chart. A sight not to be missed is a comet passing near a star cluster or other celestial object.

Though comets have for centuries been associated with evil and disaster, these celestial visitors offer excellent opportunities to observe one of nature's grandest and oldest spectacles.

38 What Are the Planetary Paths?

Making Ellipses

Level: Intermediate and advanced

Objectives: To learn about and make ellipses, the orbital shapes of nearly all celestial bodies

Materials:

- A thick piece of cardboard, about 12 inches square
- 2 pushpins
- A piece of string about 15 inches long, tied into a closed loop
- A sheet of white paper, 8 1/2 by 11 inches
- A pencil

BACKGROUND

An **ellipse** is like an oval or a lengthwise slice through a football. Ellipses belong to the family of geometric figures called **conic sections** because they are created by slicing a cone (Figure 38-1). Depending upon how the cone is cut, the figure that results may be a circle, ellipse, parabola, or hyperbola. In astronomy, all of these shapes are important in describing the orbits of planets, asteroids, and comets around the sun. The planets all orbit the sun in grand elliptical paths. In this activity, you will make ellipses and learn about some of their important astronomical applications.

Why do planets orbit the sun along elliptical paths? Elliptical orbits are really a complex result of the nature of the force of gravity. For centuries, it had been thought that the orbits of planets were circular, since this was considered the most perfect, pleasing, and harmonious of shapes. In fact, Polish astronomer Nicolaus Copernicus in 1543 stated in his book *On the Revolutions of the Celestial Spheres* that all planetary orbits are circular and that the sun is at the center of all orbits. Astronomer Tycho Brahe later set out to prove this theory. To do so, he had to carefully search for slight changes in the positions of the planets over many years. After years of painstaking work, Brahe noticed no apparent shifts in the positions of the planets among the stars and thus assumed that Copernicus's "Sun-centered solar system" idea was incorrect. In 1601, just after Brahe's death, his precise charts fell into the hands of his assistant, Johannes Kepler. After carefully analyzing all of Brahe's data, Kepler was able to prove mathematically that ellipses truly describe the orbits of planets and that the sun really did lie at the center of the solar system. In 1609, he published his results in his *New Astronomy*, in which he described his three laws of planetary motion, the first of which states that planetary paths are elliptical.

FIGURE 38-1. By slicing a cone in different ways, different geometrical shapes result at the intersection points.

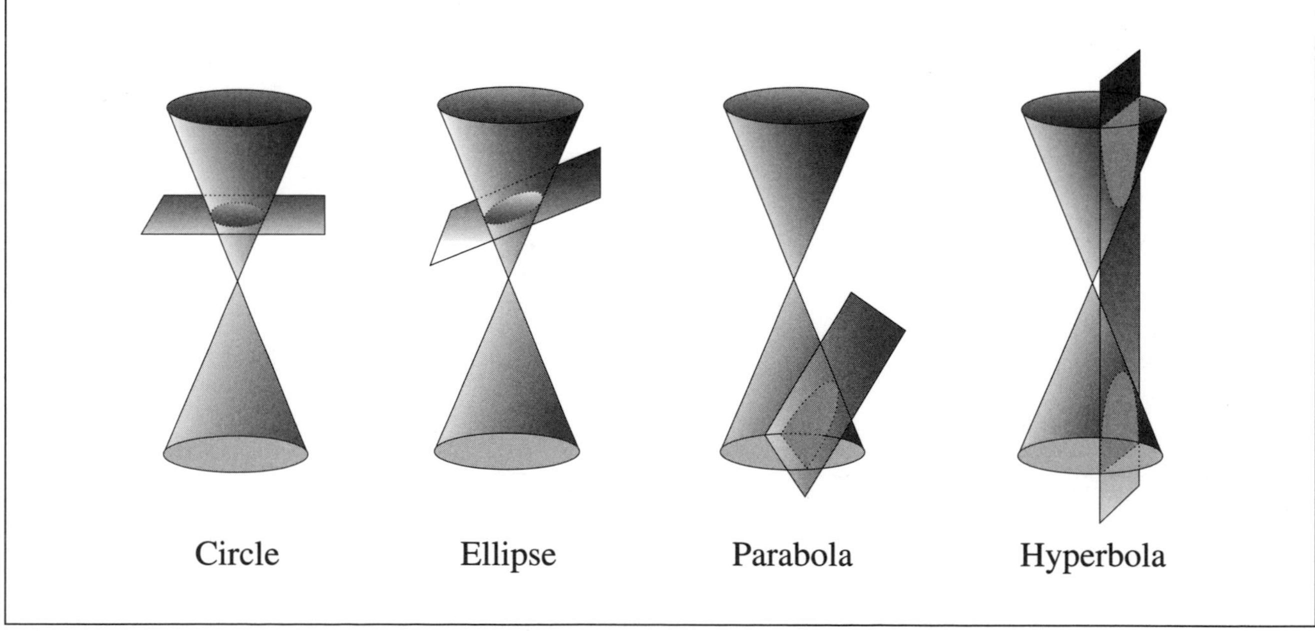

ACTIVITY

First lay down the white paper on the cardboard (as shown in Figure 38-2). Then stick the two pushpins into the cardboard near the center, about 4 inches apart. Place the loop of string around the pushpins so it lies flat on the cardboard. Place the pencil point on the inside of the string loop and, with the pencil, pull the string taut. Now move the point of the pencil around the page, keeping the loop pulled taut. You will see that you are tracing out the path of an ellipse. Continue moving the pencil, pulling the loop all the way around until the pencil returns to its starting point. To change the size of the ellipse, use different sizes of string loops or change the separation of the pushpins (just move them either closer or father apart).

After constructing some ellipses of your own, you may have a better understanding of how planets orbit the sun. Table 38-1 lists important characteristics about each planet's orbit, including the **eccentricity** of the orbit—that is, the degree of "squashiness" that the ellipse has. An ellipse of 0 eccentricity would be a circle, while an ellipse of eccentricity 0.9 would be very squashed. A fun test of your ellipse-making skills is to try to sketch ellipses with the same eccentricities as those of the true planets. Which ones are most squashed? Which ones are most circular? Making ellipses is an enjoyable way to learn more about the history of astronomy, mathematics, and the planets.

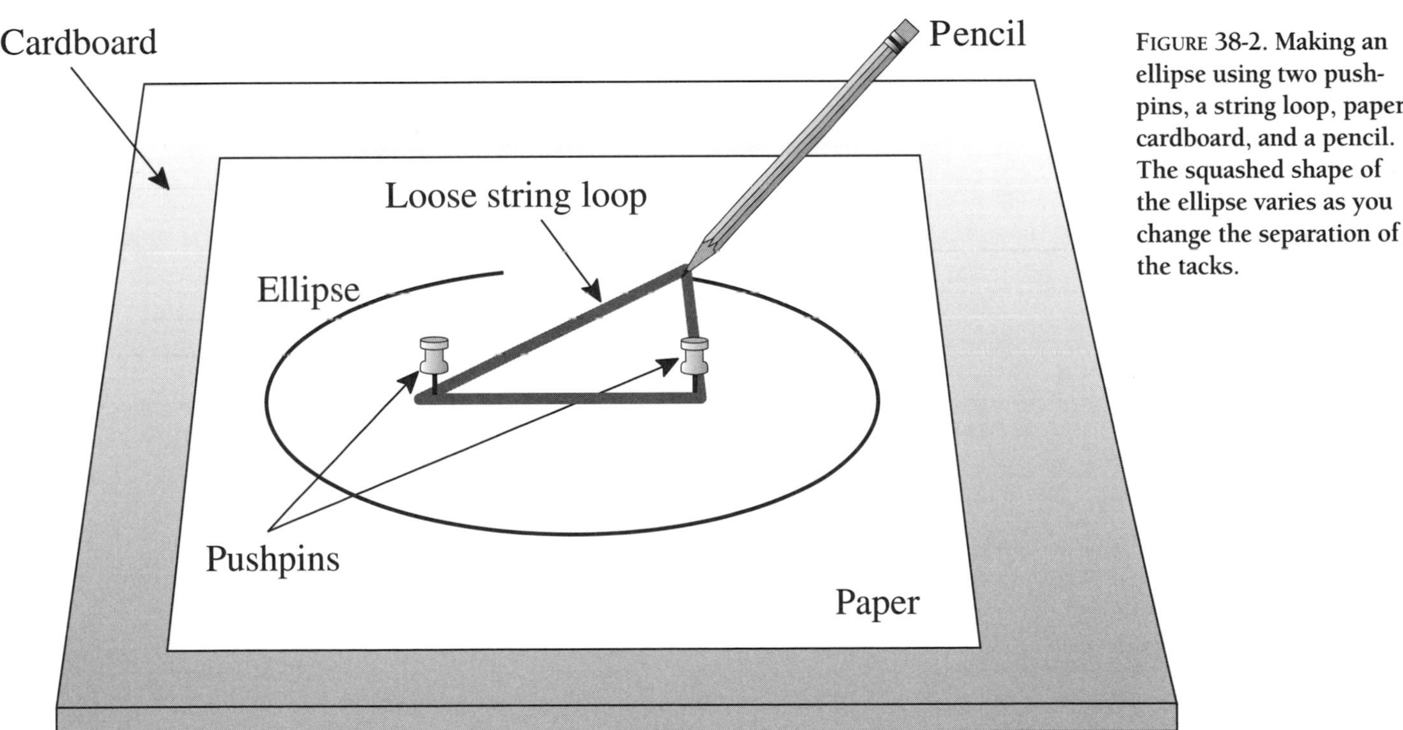

FIGURE 38-2. Making an ellipse using two push-pins, a string loop, paper, cardboard, and a pencil. The squashed shape of the ellipse varies as you change the separation of the tacks.

TABLE 38-1

Here is a list of the planets and approximate values of important properties of their elliptical orbital paths. 1 A.U. is one astronomical unit, which is 93 million miles, the average Earth-sun distance.

PLANET	AVG. DISTANCE FROM SUN (in A.U.)	PERIHELION (CLOSEST POINT TO SUN) (in A.U.)	APHELION (FARTHEST POINT FROM SUN) (in A.U.)	TIME FOR ONE ORBIT (years)	ECCENTRICITY OF ORBIT
Mercury	0.38	0.31	0.47	0.24 (88 days)	0.2
Venus	0.72	0.718	0.728	0.62 (224 days)	0.007
Earth	1.00	0.98 (around January 4)	1.02 (around July 4)	1.00 (365 days)	0.02
Mars	1.52	1.38	1.67	1.88 (687 days)	0.09
Jupiter	5.20	4.95	5.45	11.9	0.05
Saturn	9.59	9.02	10.0	29.5	0.06
Uranus	19.2	18.3	20.1	84.0	0.05
Neptune	30.1	30.0	30.3	164	0.01
Pluto	39.5	29.7	49.9	248	0.25

39 Building a Pocket Solar System

Level: Intermediate and advanced

Objective: To understand the tremendous distances in space by constructing a simple model of the solar system

Materials:

- A blank cash register tape (from a stationery store), at least 125 feet long
- Assorted Magic Markers (or crayons) and a pencil
- A tape measure
- Plenty of space (a long room or long hallway)

BACKGROUND

The solar system (or "sun's family") includes a collection of nine planets of different sizes, the largest being Jupiter, followed in size by Saturn, Uranus, Neptune, Earth, Venus, Mars, Mercury, and Pluto.

Material left over from formation of the solar system comprises a ring of rocky debris orbiting the sun between Mars and Jupiter that is known as the **Asteroid Belt**. Much of this material is similar to rocks you may see in your backyard, but with some important differences. Asteroids, unlike Earth rocks, don't experience weather (rain, snow, erosion) and for the most part have remained unchanged since their formation 4.5 billion years ago.

Our solar system also contains great numbers of icy rocks, the objects called comets (see Activity 37), believed to orbit the sun in a region called the **Oort Comet Cloud**, tremendously farther out than even tiny Pluto. Mixtures of rock and mostly water ice, comets are "dirty snowballs." Occasionally some disturbance well beyond Pluto, such as a passing star, may cause a comet to be "pushed" toward the inner solar system. Other comets, such as Halley's comet, visit predictably, periodically encountering the sun and inner planets.

In this activity, we will be working with the average distances from the sun to each of these solar system components. But we will bring these great distances "down to Earth" in a way that will help us visualize the vastness of our solar system.

Because planets do not orbit the sun in circular paths, but do so in the "squashed" circles called ellipses (see Activity 38), their actual distances from the sun are always changing. This is not to say that a particular planet can be *any* distance from the sun; its distance must fall within a certain range. For example, Earth is 94.5 million miles from the sun during early July, but only 91.3 million miles from the sun during early January. A planet's average distance from the sun is between its farthest distance (aphelion) and its closest distance (perihelion). In the case of Earth, the average is about 93 million miles. Table 39-1 provides a list of average planet-sun distances.

ACTIVITY

We will now construct a "pocket solar system," so called because it proportionately represents planetary distances on a cash register tape that you can roll up and carry with you.

For our model, we will let one astronomical unit (93 million miles) equal 1 yard (3 feet). (You could also use a different scale, such as one astronomical unit equals 1 inch. Then the model would be small enough to be hung on a wall.)

For our solar system, we will scale the planet-sun distances to this standard yardstick, as has been done in Table 39-1 in the column "Distance Between Objects on Register Tape." For example, the first entry in this column is "14 inches," indicating that Mercury must be 14 inches from the sun on your tape. Similarly, Venus must be drawn 12 inches from Mercury, and so on. As you'll see, the separations of the planets can be very large on our yardstick scale.

Even greater are the distances to the Oort Comet Cloud and to the sun's next-door neighbor, Alpha Centauri, a triple star system. Letting 1 yard equal one astronomical unit, the Oort Comet Cloud would be nearly 72 miles away, while Alpha Centauri would be 152 miles away!

To make the pocket solar system, start by unraveling about a yard of register tape. At this end, draw a large sun. Now use the distances in the "Distance Between Objects" column in

TABLE 39-1

For purposes of building a "pocket solar system," here is a list of planetary distances from the sun and from each other for the register tape. On the tape, 1 yard (3 feet) equals one astronomical unit (93 million miles).

PLANET/OBJECT	AVERAGE DISTANCE FROM THE SUN (MILES/KILOMETERS)	AVERAGE DISTANCE FROM THE SUN (IN ASTRONOMICAL UNITS)	DISTANCE FROM SUN ON REGISTER TAPE	DISTANCE BETWEEN OBJECTS ON REGISTER TAPE
Sun	–	–	–	14 inches
Mercury	36 million miles (58 million kilometers)	0.38	14 inches	12 inches
Venus	67 million miles (108 million kilometers)	0.72	26 inches	10 inches
Earth	93 million miles (150 million kilometers)	1.00	36 inches	0.1 inch
Moon	93 million miles (150 million kilometers) (Distance from Earth: 238,000 miles)	1.00	36 inches	19 inches
Mars	142 million miles (228 million kilometers)	1.52	55 inches	3.7 feet
Asteroid Belt	256 million miles (411 million kilometers)	2.75	8.25 feet	7.4 feet
Jupiter	483 million miles (778 million kilometers)	5.20	15.6 feet	13 feet
Saturn	885 million miles (1,426 million kilometers)	9.59	28.6 feet	29 feet
Uranus	1,787 million miles (2,877 million kilometers)	19.2	57.6 feet	33 feet
Neptune	2,800 million miles (4,508 million kilometers)	30.1	90.3 feet	29 feet
Pluto	3,699 million miles (5,955 million kilometers)	39.5	119 feet	71.6 miles
Oort Comet Cloud	11,700 billion miles (18,837 billion kilometers)	126,000	71.6 miles	80.5 miles
Alpha Centauri (star closest to the sun)	24,863 billion miles (40,029 billion kilometers)	268,000	152 miles	–

Table 39-1. Draw Mercury 14 inches from the sun. Next, put Venus 12 inches from Mercury, Earth 10 inches from Venus, and so on (Figure 39-1).

What happens by the time you reach Uranus? Yes, the distances between planets have become quite large. You'll just be able to get out as far as Pluto on your 125-foot-long register tape. Of course the Oort Comet Cloud and Alpha Centauri are far, far beyond the end of the tape. When you're finished drawing in the planets, you may want to find color photos of the planets and add color to your solar system.

You can now roll up the tape, keeping it ready for springing on your friends as a dramatic illustration of the tremendous scale of the solar system. Remember that it needs lots of space, such as a very long hall or a place outdoors (not on a windy day). Give the "sun" end of the roll to someone to hold as you unroll the tape. You might explain which planets are moving past, noting that the temperature and sunlight drop markedly as you approach the outer solar system. (Near Pluto the temperature drops to 425° F below zero.) To entice listeners into a discussion of the plausibility of space travel, make note of the distances to the Oort Comet Cloud and to the nearest star system—and throw in the fact that a space shuttle ride to Alpha Centauri would take 125,000 years!

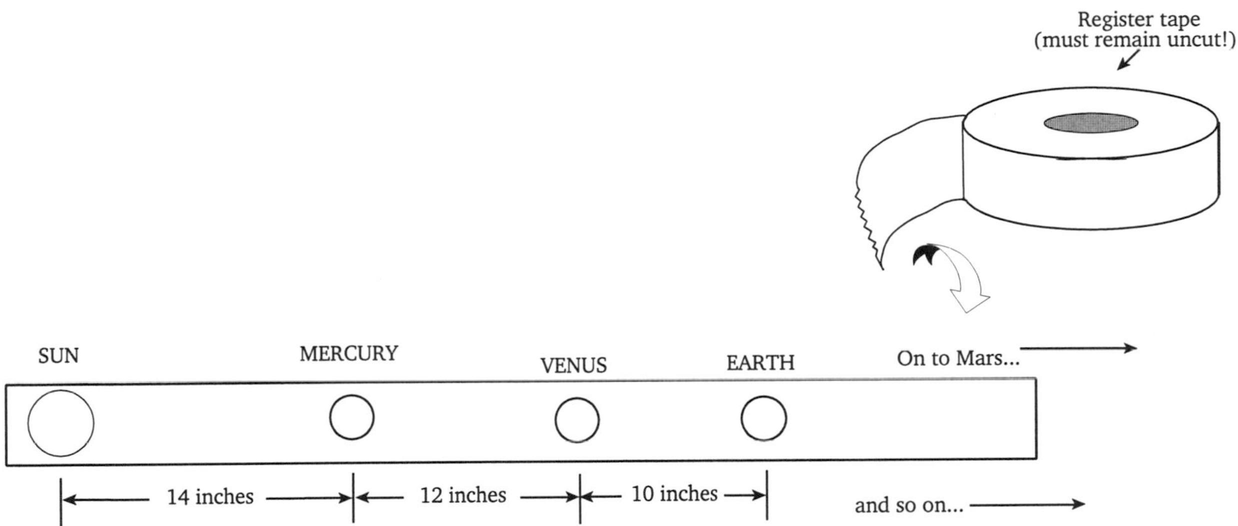

FIGURE 39-1. The Pocket Solar System should begin as shown, with the distance between each planet marked as the tape is unrolled. Note that only the relative distances of the planets can be shown, not their relative sizes.

SECTION V

Deep Space

40 Seeing Double

Observing Double Stars

Level: Intermediate and advanced

Objectives: To observe how many of the stars at night are actually accompanied by fainter companion stars

Materials:

- A telescope or binoculars
- A clear evening sky
- The appropriate seasonal star map (from Activities 4–7)
- Photocopies of Double Star Observation Form (included with this activity)

BACKGROUND

Take a look up at the night sky. What do you see? Stars, yes, but what do they look like? They look like single points of light, right? Not all stars are just **single** points, however. Astronomers now believe that as many as half of them are actually **double stars** or **multiple stars**. Double and multiple stars are defined as one or more stars orbiting another star, similar to how Earth and the other planets orbit the sun. But some of these are impostors, are really just two stars that happen to lie along the same line of sight as seen from Earth. These stellar pretenders are called **optical doubles**.

FIGURE 40-1. The beautiful double star Albireo, found in the summer constellation Cygnus the Swan.

ACTIVITY

Take a look toward the Big Dipper and its home constellation Ursa Major. Turn your attention toward the middle star in the handle of the Dipper, the star named Mizar. If you have keen eyesight, you just might be able to make out a second, fainter star right next door. This second star is named Alcor. Alcor and Mizar make up the best known naked-eye double star in the sky. But studies of these stars show that they are actually nowhere near each other in space. They just happen to lie along the same line of sight. In reality, Alcor and Mizar form an optical double.

If you have a telescope, take a close look at the brighter of those two stars, Mizar. At about 25**x** or more, you might be able to tell that Mizar is actually a double star in its own right. Look carefully for two equally bright points of white light almost touching each other. Based on studies conducted over the past century or more, astronomers have concluded that these two stars are actually bound to each other gravitationally. Mizar, therefore, is a true double star.

The separation between individual members of a double star system as seen from Earth is usually expressed as **angular separation** (see Activity 13), measured in minutes or seconds of arc. One degree (1°) of angle equals 60 arc-minutes (60') , and each arc-minute (1') equals 60 arc-seconds (60"). (Arc-minutes and arc-seconds are angular measurements and have nothing to do with minutes and seconds of time.) The stars may actually be hundreds of millions, even billions, of miles apart.

TABLE 40-1

This table lists the finest double stars visible from mid-northern latitudes. Broken down by season, each entry lists the star's name or designation and its home constellation, as well as the magnitudes of the stars involved, their separation, and any color. Also noted is the smallest optical instrument needed to see the double star, either a telescope or binoculars. This way, if you see a star on the list that needs, say, a 2-inch telescope to show the components, and you have a 3-inch telescope, then you ought to be able to make them out. If, on the other hand, all that you have is a pair of binoculars, bypass that star in favor of another selection.

The Sky's Best Double Stars

STAR	CONSTELLATION	RIGHT ASCENSION		DECLINATION		MAG.	SEP.	COLORS	TELESCOPE REQUIRED
Spring		h	m	°	'				
Cor Caroli	Canes Venatici	12	56.0	+38	19	3,6	20"	White, white	3-inch
Alcor & Mizar	Ursa Major	13	23.9	+54	56	2,4	12'	White, white	Binoculars
Mizar	Ursa Major	13	23.9	+54	56	2,4	14"	White, white	3-inch
Summer									
Rasalgethi	Hercules	17	14.6	+14	23	3,6	4"	Orange, green	6-inch
Albireo	Cygnus	19	30.7	+27	58	3,5	35"	Yellow, blue	2-inch
Autumn									
Algedi	Capricornus	20	21.0	–14	47	3,6	3'	Yellow, white	Binoculars
Almach	Andromeda	02	03.9	+42	20	2,5	10"	Orange, blue	3-inch
Winter									
Rigel	Orion	05	14.5	–08	12	0,7	9"	Blue, white	4-inch
Pollux	Gemini	07	34.6	+31	53	2,3	3"	White, white	6-inch

Hints for finding these stars: All of the stars listed in this table are shown on their respective seasonal star charts found near the beginning of this book.

Note: Right ascension and declination are coordinates astronomers use to mark the location of a celestial body. Think of them as a sort of astronomical longitude and latitude.

Double Star Observation

Star: ______________________ Constellation: ______________________

Date: ______________________ Time: ______________________

Telescope: ____

Eyepiece: ______________________ Power: ______________________

Notes:

1. Do the stars display any distinctive color? If so, what? ______________________

2. How difficult is it to split the double star into its separate components?

☐ Easy ☐ Moderately difficult ☐ Impossible

3. Other comments: ____

____

____

Each season the sky holds many fascinating and beautiful double stars (also called binary stars) in store for stargazers with telescopes and binoculars. Many of these stars, such as the one shown in Figure 40-1, display striking contrasts in magnitude, while others are nearly equal in brightness. Some seem to shine pure white, while others glimmer with distinctive colors such as blue, yellow, red, and orange.

On the next clear night, go outside with your telescope or binoculars and try to find as many of the double stars in Table 40-1 as possible. First adjust your telescope's magnification (if possible) until you see the stars' images crisp and clear. Use just enough magnification to spot the stars, but not so much as to make them unfocusable. Look carefully to see if you can spot any color in the stars. If not, try defocusing the star images just slightly. Sometimes that will make subtle colors a little more obvious.

Using a soft pencil like a #2, draw what you see onto a copy of the Double Star Observation Form. Include not only the double star but other stars in the field of view as well. Try to draw the view as accurately as possible. Show the brighter stars as larger dots than the fainter ones. Make a quick sketch at the telescope, then go inside and finish it off.

By the time you are through with this activity, you should have a set of drawings of some of the finest double stars the sky has to offer.

41 Families of Stars

Observing Star Clusters

Level: Intermediate and advanced

Objectives:

- To learn about how some stars belong to large, extended families called star clusters
- To observe examples of star clusters

Materials:

- A clear, dark sky far from bright lights
- Binoculars or a small telescope
- A red-filtered flashlight
- Detailed Star charts (pages 202–206)
- Photocopies of Open Cluster Observation Form and Globular Cluster Observation Form (included with this activity)

BACKGROUND

By now it should be clear that there is a lot more to this universe of ours than meets the eye. Many of the night sky's finest treasures lie buried among the stars, visible only with the help of a pair of binoculars or a telescope. As already discussed in Activity 40, on double stars, some seemingly mundane stars reveal striking personalities when studied with optical instruments. If you thought those were interesting, "you ain't seen nothin' yet!"

The activity on double stars noted that as many as half of the stars seen at night are actually close-set stellar pairs. As many as half a dozen or more stars may be clumped together in multiple star systems. Sprinkled throughout and around our galaxy, the Milky Way, are huge "extended families" of stars, with dozens, hundreds, thousands, and even more individual suns held together under the force of gravity. Astronomers call such a huge number of stars collected in a relatively small volume of space a **star cluster**.

Star clusters come in two basic varieties: **open star clusters** and **globular star clusters**. Open clusters, also referred to as galactic clusters, are by far the most plentiful. Each is a randomly shaped family of mostly young, hot blue-white stars, as shown in Figures 41-1, 41-2, and 41-3. An open cluster may contain anywhere from a few to a few hundred individual points of light. All of the stars in a given cluster lie about the same distance from Earth and are all about the same age, having formed from a common cloud of interstellar gas and dust called a **nebula** (discussed in the next activity). The stars in an open cluster travel through space together, only to scatter slowly over eons of time.

Figure 41-1. All of the bright stars in the center of this photograph belong to the open star cluster M44 in the spring constellation Cancer. (Photo by George Viscome)

Figure 41-2. Open star cluster M11 in the faint summer constellation Scutum, found just south of Aquila. (Photo by George Viscome)

FIGURE 41-3. The double cluster, a pair of open star clusters in the autumn constellation Perseus. (Photo by George Viscome)

The second type of cluster, globular clusters, is shown in Figures 41-4 and 41-5. Globular clusters contain many times the number of stars found in open clusters, with populations usually in the hundreds of thousands. Unlike open clusters, which are found scattered inside the boundaries of the Milky Way, globular clusters surround our galaxy much as moths surround a porch light or a flame. Each globular cluster is composed of ancient stars, all much older than our 4.5-billion-year-old sun.

FIGURE 41-4. Globular star cluster M5 in the spring constellation Serpens. (Photo by George Viscome)

ACTIVITY

Many open clusters and globular clusters are bright enough to be seen in binoculars and small telescopes. Some are even visible to the naked eye on exceptionally clear and dark nights. Table 41-1 lists some of the finest examples of each, broken down by seasonal sky. Using the detailed star charts (finder charts) on pages 202–206, see how many of each you can find.

Open Cluster Observation

Object: ______________________ Constellation: ______________________

Date: ______________________ Time: ______________________

Telescope: ____

Eyepiece: ______________________ Power: ______________________

Notes:

1. About how many individual stars can you see? ______________________

2. Do any of the stars have a distinctive color? If so, what? ______________________

3. What is the overall shape of the cluster? ______________________

4. Playing connect-the-dots, does the cluster's pattern resemble anything you have seen on Earth? If so, what? ______________________

5. Other comments: ____

____

Globular Cluster Observation

Object: ____________________ Constellation: ____________________

Date: ____________________ Time: ____________________

Telescope: ____

Eyepiece: ____________________ Power: ____________________

Notes:

1. Can you see any of the cluster's individual stars? (Hint: Look around the edges) ____________
2. What is the cluster's shape? Is it round or oval? ____________________
3. Can you see any color to the cluster? If so, what? ____________________
4. Other comments: ____

____

____

TABLE 41-1

The Sky's Best Star Clusters

NICKNAME	CATALOG LISTING [1]	CONSTELLATION	RIGHT ASCENSION	DECLINATION	MAGNITUDE	TELESCOPE REQUIRED
		OPEN CLUSTERS				
Spring			h m	° '		
Praesepe Cluster	M44	Cancer	08 40.1	+19 59	3	Binoculars
	M67	Cancer	08 50.4	+11 49	7	2-inch
Coma Berenices	Cluster	Coma Berenices	12 25.1	+26 06	3	Binoculars
Summer						
Butterfly Cluster	M6	Scorpius	17 40.1	–32 13	4	Binoculars
	M7	Scorpius	17 53.9	–34 49	3	Binoculars
Wild Duck Cluster	M11	Scutum	18 51.1	–06 16	6	2-inch
Coathanger	—	Vulpecula	19 25.4	+20 11	4	Binoculars
Autumn						
Double Cluster	NGC 869	Perseus	02 19.0	+57 09	4	Binoculars
	NGC 884	Perseus	02 22.4	+57 07	4	Binoculars
Perseus Cluster		Perseus	03 22	+48 36	2	Binoculars
Winter						
Pleiades	M45	Taurus	03 47.0	+24 07	1	Binoculars
Hyades	—	Taurus	04 27	+16	1	Naked eye
	M37	Auriga	05 52.4	+32 33	6	2-inch
	M35	Gemini	06 08.9	+24 20	5	2-inch
		GLOBULAR CLUSTERS				
Spring			h m	° '		
	M3	Canes Venatici				2-inch
	M5	Serpens	15 18.6	+02 05	6	2-inch
Summer						
	M4	Scorpius	16 23.6	–26 32	6	2-inch
	M13	Hercules	16 41.7	+36 28	6	Binoculars
	M22	Sagittarius	18 36.4	–23 54	5	Binoculars
Autumn						
	M15	Pegasus	21 30.0	+12 10	6	2-inch
	M2	Aquarius	21 33.5	–00 49	7	2-inch

[1] Astronomers have assembled different ways of cataloging objects in the sky. The M designation indicates the object referenced is in the Messier catalog of star clusters, nebulae, and galaxies, while NGC refers to the New General Catalog. M2, for example, indicates this to be the second entry in the Messier catalog. There are 109 members of the Messier catalog and over 7,800 objects in the NGC.

Note: Right ascension and declination are coordinates astronomers use to mark the location of celestial objects. Think of them as a sort of astronomical longitude and latitude.

FIGURE 41-5. Globular star cluster M22 in the summer constellation Sagittarius. (Photo by George Viscome)

As you are looking at each star cluster, write down your impressions on the appropriate Cluster Observation Form. Draw the cluster as best you are able, noting any features mentioned in your written description. By studying stars in both open and globular clusters, astronomers expand their knowledge not only of the universe, but of our own star, the sun, as well.

42 Clouds in Space

Observing Nebulae

Level: Intermediate and advanced

Objectives:

- To learn about regions where stars are born and die
- To observe examples of stellar formation and stellar death

Materials:

- A clear, dark sky far from bright lights
- Binoculars or a small telescope
- A red-filtered flashlight
- Detailed star charts (see pages 202–206)
- Photocopies of Nebula Observation Form (included with this activity)

BACKGROUND

Not all of the clouds seen in the sky are found in Earth's atmosphere. Using telescopes, astronomers over the past three centuries have discovered distant clouds in the night sky that are not only located beyond Earth but even beyond the edge of our solar system. These clouds are found out among the stars. They are not clouds of water vapor, like those in our atmosphere, but instead are composed of a complex mix of dust and hydrogen, helium, nitrogen, and oxygen gases. So as not to confuse them with the clouds in our sky, astronomers call these interstellar clouds **nebulae** (plural of **nebula**). Astronomers now know that there are two broad categories of nebulae found in the sky.

Do you know where stars are born? No, not in Hollywood! Real stars are born in **diffuse nebulae**. The word "diffuse" means "scattered or spread out, not concentrated." As the accompanying photographs attest, this is an apt description. Diffuse nebulae are formless blobs of cloudiness that have no exact shape or form. Their great distance from us makes them appear very small in our sky, but in reality, diffuse nebulae are much larger than our entire solar system. In fact, it is thought that the sun and all of its planets (including Earth) were formed from a single diffuse nebula about 4.5 billion years ago. Figures 42-1 and 42-2 show two outstanding diffuse nebulae.

Across an immeasurable period of time, the molecules and dust particles in this nebula collided with each other and, drawn together by gravity, slowly gained energy. Almost leisurely, the cloud started to spin. As it rotated, the cloud slowly pulled itself inward.

FIGURE 42-1. (above) The diffuse nebula M42, the famous Orion Nebula, highlights the center of this photograph of Orion's sword. (Photo by George Viscome)

FIGURE 42-2. (left) Diffuse nebula M17, the Horseshoe Nebula, in the summer constellation Sagittarius. (Photo by George Viscome)

Just as figure skaters spin faster as they draw their arms in, the cloud picked up speed as more and more of its material fell in toward its center. As the accumulation of gases increased, the cloud's temperature began to rise, first slightly, then rapidly. Eventually, with most of the cloud's material pulled toward its center, the temperature was sufficient for the infant sun to "turn on."

As the early sun was forming, other, smaller pockets of cloudiness were condensing within the nebula as well. Though these eddies were not large enough to create another star, they ultimately produced our planets. The remains of the cloud itself has long since dissipated, leaving behind the solar system we know today.

Just as a diffuse nebula marks the birthplace of new stars, a second type of nebula denotes the grave of old stars. Examples of this second variety are called **planetary nebulae**. Don't be fooled by the name; planetary nebulae have nothing to do with planets. The name comes from the fact that through most telescopes, planetary nebulae look like tiny, round disks of light, at first glance just like a small planet. In reality, however, planetary nebulae are also located well beyond our solar system.

A star is primarily made of hydrogen and spends most of its life converting this hydrogen to helium and heavier elements in a complex thermonuclear reaction. Though a star contains an incredible amount of fuel, this resource is not inexhaustible. (It is predicted that the sun will run out of fuel in about 4.5 billion years.) When a star has used up its fuel, drastic changes commence. The star will swell to immense proportions, cooling in the process. Reaching a maximum diameter (which can vary dramatically depending on how large it was to begin with), it then begins to shrink and rise in temperature. In the process a cloud of gaseous material will be discharged. As the star continues to lose energy, the gaseous "shell" will expand to surround the dying star. Over the course of millions of years, the star will fade away and the cloud disperse, but in the meantime we see this shell of gas in the form of a planetary nebula, such as those seen in Figures 42-3 and 42-4.

FIGURE 42-3. (left) The famous Ring Nebula M57, a planetary nebula in the summer constellation Lyra. (Photo by George Viscome)

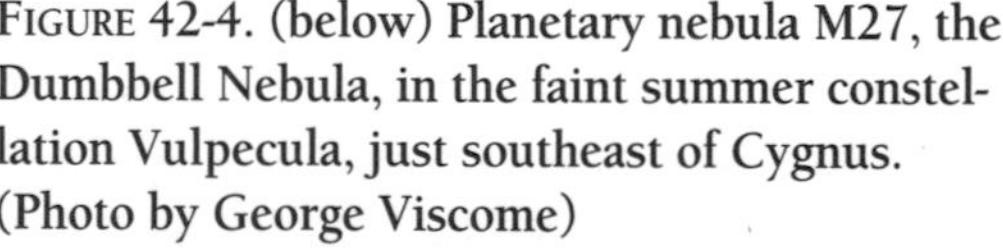

FIGURE 42-4. (below) Planetary nebula M27, the Dumbbell Nebula, in the faint summer constellation Vulpecula, just southeast of Cygnus. (Photo by George Viscome)

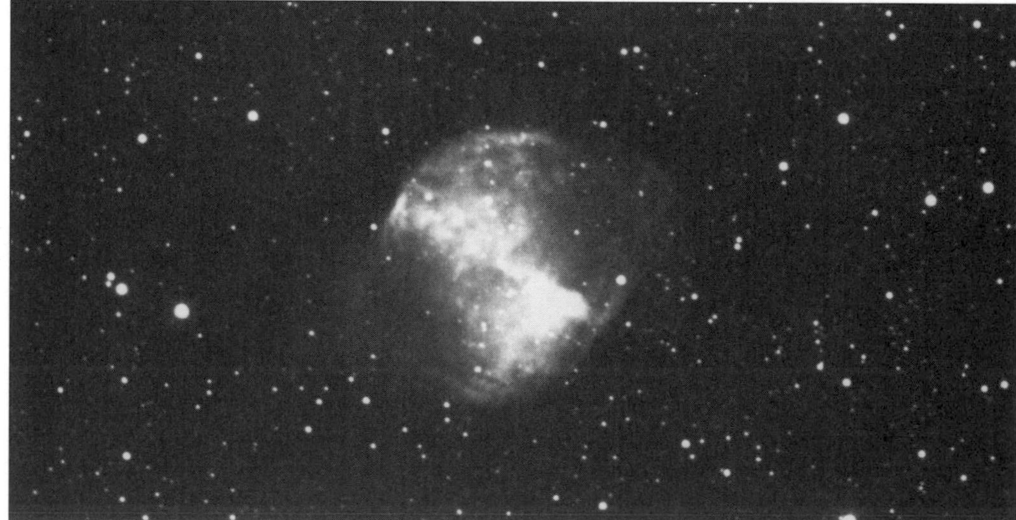

ACTIVITY

Many examples of diffuse and planetary nebulae are found throughout the sky, with the brighter examples situated in the summer and winter skies. Table 42-1 is a list of the finest nebulae visible through binoculars and small telescopes. Be forewarned, however, that these objects will *not* appear as they do in photographs. Pictures are capable of revealing much finer and subtler detail than our eyes can. Instead of seeing magnificent, colorful swirls, chances are all you will see is a grayish smudge of faint light. But while the visual impact may at first seem a little disappointing, it is important to remember *what* you are looking at. You are seeing, firsthand, a wondrous example of how our universe works from across a void of hundreds, even thousands, of light years.

TABLE 42-1

The Sky's Best Nebulae

NICKNAME	CATALOG LISTING [1]	CONSTELLATION	RIGHT ASCENSION	DECLINATION	MAGNITUDE	TELESCOPE REQUIRED
		DIFFUSE NEBULAE				
Summer			h m	° '		
Trifid Nebula	M20	Sagittarius	18 02.6	–23 02	9	3-inch
Lagoon Nebula	M8	Sagittarius	18 03.8	–24 23	6	Binoculars
Omega Nebula	M17	Sagittarius	18 20.8	–16 11	7	2-inch
Winter						
Crab Nebula	M1	Taurus	05 34.5	+22 01	8	3-inch
Orion Nebula	M42	Orion	05 35.4	–05 27	4	Binoculars
	M78	Orion	05 46.7	+00 03	8	3-inch
		PLANETARY NEBULAE				
Summer						
Ring Nebula	M57	Lyra	18 53.6	+33 02	9	3-inch
Dumbbell Nebula	M27	Vulpecula	19 59.6	+22 43	8	2-inch
Winter						
Eskimo Nebula	NGC 2392	Gemini	07 29.2	+20 55	8	4-inch

1. Astronomers have assembled different ways of cataloging objects in the sky. The M designation indicates the object referenced is in the Messier catalog of star clusters, nebulae, and galaxies, while NGC refers to the New General Catalog. M42, for example, indicates this to be the forty-second entry in the Messier catalog. There are 109 members of the Messier catalog and over 7,800 objects in the NGC. **Note:** Right ascension and declination are coordinates astronomers use to mark the location of celestial objects. Think of them as a sort of astronomical longitude and latitude.

All of the nebulae listed in Table 42-1 are plotted on the detailed star charts (finder charts) on pages 202–206. Compare these charts to the seasonal star maps first introduced in Activities 3 through 7. By going back and forth among the appropriate seasonal star map, the finder chart, and your telescope or binoculars, the nebulae will reveal themselves.

Nebula Observation

Object: ______________________ Constellation: ______________________

Date: ______________________ Time: ______________________

Telescope: ____

Eyepiece: ______________________ Power: ______________________

Notes:

1. What type of nebula is the object? ☐ Diffuse ☐ Planetary

2. Do you see any distinct shape to the nebula? If so, what? ______________________

3. Can you see any stars in the nebula? If so, how many? ______________________

4. What is the color of the nebula? ______________________

5. Other comments: ______________________

____

____

For each nebula spotted, complete a Nebula Observation Form. Answer as many of the questions as you can. If another person is viewing along with you, record his or her impressions as well. It is always interesting to see how two people's views of the same object can differ greatly.

The most important thing to remember when looking for nebulae (or any other sky object, for that matter) is not to rush. Some of these objects can be very difficult to see at first. If you are scanning back and forth too quickly, the intended target may be skimmed right over without your ever noticing it.

If you cannot find an object at first, take a deep breath and try this trick. Instead of staring directly at the target area, try looking a little to one side of the eyepiece's field of view. By averting your vision, light falls onto a more sensitive part of the eye's retina, making it possible to glimpse dim objects that are otherwise invisible with direct vision. If an object eludes you even with averted vision, move on to another object and return later. Even when spotted, however, the nebulae will not look like the photographs shown here, but instead only as dim smudges of gray light. Still, there is nothing like the thrill of actually seeing the universe through your own telescope. Be sure to give each nebula a try.

43 Distant Islands of Stars

Observing Galaxies

Level: Advanced

Objectives:

- To locate and view examples of galaxies beyond our own
- To understand the structure of our galaxy, the Milky Way, and where we fit into the universe

Materials:

- A clear, dark sky far from bright lights
- Binoculars or a small telescope
- A red-filtered flashlight
- Detailed star charts (see pages 202–206)
- Photocopies of Galaxy Observation Form (included with this activity)

BACKGROUND

Until now, everything we have explored in this book—the moon, the sun, all of the planets in the solar system, stars and star clusters, and nebulae—is found within the single huge complex called the Milky Way. The Milky Way is our galaxy, a collection of about 300 billion to 400 billion stars. Everything we see in the sky belongs to the Milky Way.

Looking beyond the Milky Way, astronomers can see many other galaxies in the emptiness of the universe, like islands in a vast ocean . It is now known that the Milky Way is but one of billions of galaxies. When we gaze toward other galaxies, our view is stretching across tremendous distances. The closest major galaxy to our own is called the **Andromeda Galaxy** (Figure 43-1) and is believed to lie some 2.2 million light-years (or about 13 quintillion miles) away. And that's the closest!

Figure 43-1. The spiral galaxy M31 in the autumn constellation Andromeda. Also note the two smaller galaxies on either side of M31. Below is M32, while above is M110. Both of these elliptical galaxies are gravitationally bound to larger M31. M32 is visible in 3-inch telescopes, while fainter M110 requires at least a 4-inch telescope to be seen. (Photo by George Viscome)

It is difficult to imagine what the Milky Way looks like from the outside, since we find ourselves stuck on the inside. But by investigating other galaxies and comparing the findings with what we can see of the Milky Way, astronomers can begin to piece together the puzzle of what our "island" of stars looks like.

From studies conducted in the early part of the twentieth century, astronomers know there are three types of galaxies in the universe.

Spiral galaxies (such as shown in Figures 43-1, 43-2, and 43-3) are shaped like enormous pinwheels, with two or more spiral arms winding away from the nucleus, or center, of the galaxy. The Milky Way is believed to be a spiral galaxy.

Elliptical galaxies are the most common galaxies in the universe. Two examples are seen adjacent to the Andromeda Galaxy in Figure 43-1. Rather than showing a spiral shape, elliptical galaxies appear as enormous spheres of stars. Some elliptical galaxies appear perfectly round, while others are noticeably oval.

Irregular galaxies constitute the third class of galaxies. Irregular galaxies, as the name implies, do not fit into either of the other two categories and may be thought of as the "mongrels" of the universe. These include unusually shaped galaxies, colliding galaxies, and galaxies undergoing violent energy outbursts.

FIGURE 43-2. The spiral galaxy M81 in the circumpolar constellation Ursa Major. (Photo by George Viscome)

FIGURE 43-3. The spiral galaxy M51 in the spring constellation Canes Venatici, a small, faint constellation just below the handle of the Big Dipper. (Photo by George Viscome)

ACTIVITY

Though they lie at almost unimaginable distances, many galaxies are visible through binoculars and small telescopes *if you know exactly where to look*. Through smaller instruments, none of the galaxies appear as they do in the accompanying photographs; these were all taken through larger instruments with special film and other accessories. Instead, backyard telescopes reveal galaxies as softly glowing patches of light frequently highlighted by a bright central core. Their shapes vary from perfectly circular to cigar-shaped.

Trying to find even the brightest galaxies is one of the most challenging tasks for a stargazer. The brightest galaxy visible from the northern hemisphere, the Andromeda Galaxy, is barely perceptible to the naked eye; all others require at least a good pair of binoculars to be seen.

All of the galaxies listed in Table 43-1 are plotted on the detailed star charts (finder charts) found on pages 202–206. As with the star clusters and nebulae from the two previous activities, these charts should be compared with the seasonal star maps first introduced in Activities 3 through 7. By going back and forth between the appropriate seasonal star map, the finder chart, and your telescope or binoculars, the galaxies will reveal themselves.

Once again, you should not expect the galaxies to appear through your telescope as they do in the pictures here. These photographs are the result of very long time exposures taken through a large telescope. Through your telescope, each galaxy will probably appear, at least at first, as a faint amorphous glow. But with time you might be able to pick out some features. To help guide you along, complete a copy of the Galaxy Observation Form for each new galaxy you find. Try to answer each of the form's questions. There are no right or wrong answers; it's what you see that counts.

TABLE 43-1

The Sky's Best Galaxies

CATALOG LISTING [1]	CONSTELLATION	RIGHT ASCENSION [4]	DECLINATION[4]	MAGNITUDE	TYPE [2]	TELESCOPE REQUIRED [3]
Spring		h m	° ′			
M81	Ursa Major	09 55.6	+69 04	7	S	3-inch
M82	Ursa Major	09 55.8	+69 41	8	I	3-inch
M65	Leo	11 18.9	+13 05	9	S	4-inch
M66	Leo	11 20.2	+12 59	9	S	4-inch
M106	Canes Venatici	12 19.0	+47 18	8	S	3-inch
M94	Canes Venatici	12 50.9	+41 07	8	S	3-inch
M51	Canes Venatici	13 29.9	+47 12	8	S	3-inch
Autumn						
NGC 7331	Pegasus	22 37.1	+34 25	10	S	6-inch
M31	Andromeda	00 42.7	+41 16	4	S	Binoculars
M32	Andromeda	00 42.7	+40 52	8	E	3-inch

Notes:

[1]Astronomers have assembled different ways of cataloging objects in the sky. The M designation indicates the object referenced is in the Messier catalog of star clusters, nebulae, and galaxies, while NGC refers to the New General Catalog. M32, for example, indicates this to be the thirty-second entry in the Messier catalog. There are 109 members of the Messier catalog and over 7,800 objects in the NGC.

[2]Type of galaxy: S = spiral, E = elliptical, I = irregular

[3]This is assuming suburban sky conditions. Darker, rural skies may permit spotting these objects in smaller instruments.

[4]Right ascension and declination are coordinates astronomers use to mark the location of celestial objects. They can most simply be understood as astronomical longitude and latitude.

Galaxy-hunting is one of the most difficult aspects of telescopic astronomy. If you cannot find them immediately, have patience; they are out there. With time, all of the galaxies in Table 43-1, and many others, will reveal their dim, distant glimmer. While your neighbors watch reruns on television, you will be seeing ancient starlight from halfway across the universe!

Galaxy Observation

Object: ____________________ Constellation: ____________________

Date: ____________________ Time: ____________________

Telescope: ____

Eyepiece: ____________________ Power: ____________________

Notes:

1. Does the galaxy have a distinct shape? If so, what? ____________________

2. Does the galaxy appear uniformly bright, or are some parts brighter than others? ____________________

____

3. Can you make out a bright middle in the galaxy? ____________________

5. Other comments: ____________________

____

____

SECTION VI

Telescopes & Photography

44 Fun with Optics

Level: Intermediate and advanced

Objectives:

- To learn some basic terms about telescopes and optics
- To understand what is meant by the focal length, aperture, and focal ratio of a lens or mirror

Materials:

- A magnifying glass
- Magnifying mirror (like those used for applying makeup)
- A yardstick or tape measure
- A bright light

BACKGROUND

If you have ever looked through a telescope, you know what a great difference this instrument can make in looking at far-off sights. A tiny object suddenly looks bigger because a telescope magnifies the image. To do this, some telescopes use lenses, others use mirrors, while still others use both.

To understand how a telescope works, it is important to understand how lenses and mirrors work. Lenses and mirrors, along with prisms, may be grouped under the broad heading of **optics**.

Can you think of everyday ways in which optics influences your life? You're able to read this book thanks to optics you were born with. Each of your two eyes has a built-in lens, allowing you to focus on whatever you're looking at. If you wear glasses, your vision is improved because of other optics.

Eyeglasses use two lenses, one in front of each eye, to change the direction of light entering your eyes, helping them focus on whatever you are looking at.

Any other ways? Go into your bathroom or bedroom and look into the mirror. Most mirrors are flat, causing them to reflect an image straight back, allowing you to see yourself. If you have ever walked through a fun house at a carnival, you may have seen what can happen when a mirror is not flat. If the mirror bows outward toward you (what opticians call **convex**), the reflected image will appear smaller than it really is; if the mirror bows inward, or away from you (**concave**), the reflected image appears larger. Some small mirrors used for putting on makeup are slightly concave.

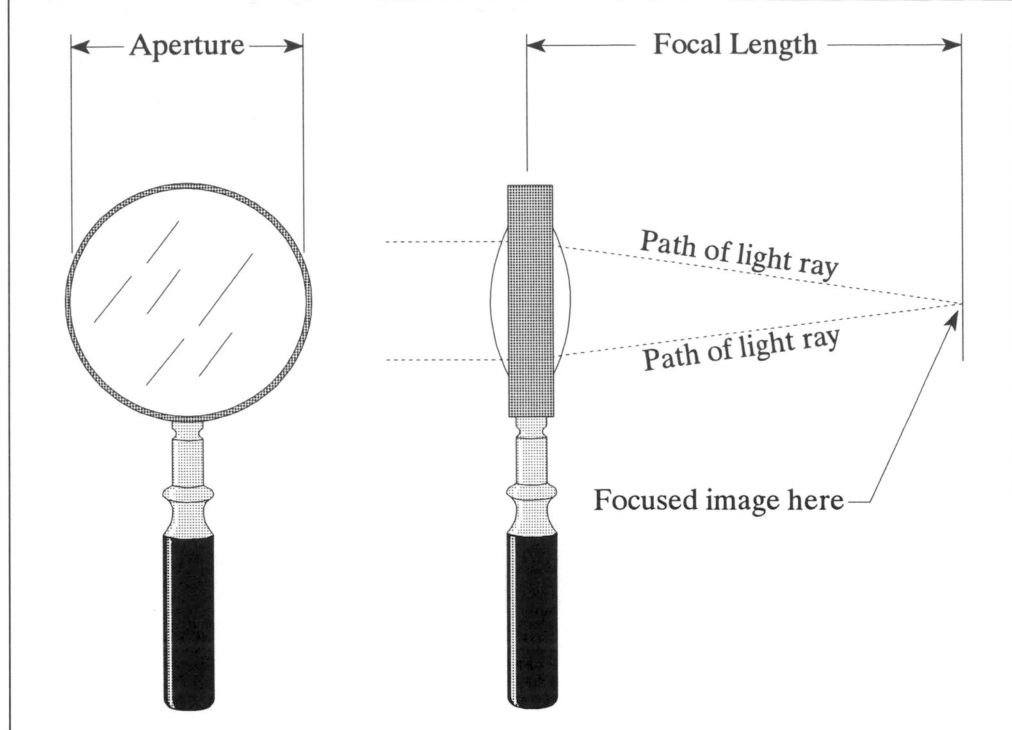

FIGURE 44-1. Diagram identifying a magnifying glass's aperture and focal length. A mirror is similar.

Do you own a magnifying glass? Magnifying glasses (Figure 44-1) are also a type of optics, in this case a lens. Take a look through the magnifying glass. It will make everything appear bigger. Turn the lens sideways and take a look at its shape. Notice how the lens bends outward on both sides. Opticians call this a **double-convex** lens. The lens of the human eye is a double-convex lens. Convex lenses make things look bigger, while concave lenses make things look smaller—just the opposite of mirrors.

ACTIVITY

Turn the magnifying glass so that you are looking straight through it. Using the yardstick or tape measure, find the diameter of the magnifying glass itself (not including any plastic used to hold the lens). The result is called the magnifying glass's **aperture**. Write this number down in Table 44-1; we will come back to it in a minute.

Notice that when you look through a magnifying glass, the view is made sharp and clear by moving the lens closer or farther from whatever you are looking at. This process is called focusing. The distance it takes light passing through the magnifying glass to come to a focus is called the lens's **focal length**.

Switch on a light source. It might be a light in your kitchen, bedroom, or living room. Place a piece of paper under the light and hold the magnifying glass somewhere in between. Notice how you can see an image of the light source on the piece of paper. By moving the lens closer to and farther from the paper, find the distance where the projected image is sharpest. That distance from the paper to the lens when the image is sharpest is the lens's focal length. Use the yardstick or tape measure to measure this distance and write it down on the chart as well.

TABLE 44-1

Magnifying Glass Facts and Figures

	APERTURE	FOCAL LENGTH	FOCAL RATIO (FOCAL LENGTH DIVIDED BY APERTURE)
Magnifying Glass	______	______	______
Concave Mirror	______	______	______

We now know two things about the magnifying glass: its aperture and its focal length. By knowing these two numbers, we can now figure out the lens's **focal ratio** or **f-number**. The focal ratio is nothing more than a lens's (or mirror's) focal length divided by its aperture. All camera lenses and telescopes also refer to an f-number, such as f/4 or f/8.

To find out the focal ratio of the magnifying glass, divide the measured focal length by the aperture. What do you get? Well, if, for instance, the magnifying glass measures 2 inches across (aperture) and the focal length measures 8 inches, then the focal ratio is 4 (written f/4), since 8 divided by 2 equals 4.

If you or someone in your family has a **concave** mirror, such as one used to apply makeup, try the same experiment. First measure the aperture, then the focal length. To measure the mirror's focal length, reflect the image of a room light off the mirror and onto a wall or piece of paper. Move the mirror back and forth until the reflection of the light is sharp and focused. Then measure the distance between the mirror and the paper or wall. Finally, figure out the mirror's focal ratio. Write the results into the table. Which has the shorter, or "faster," focal ratio, the magnifying glass or the concave mirror?

This is just a brief introduction to the fantastic world of optics. The next activity shows how all this relates to telescopes.

45 A Telescope and Binocular Primer

Level: Intermediate and advanced

Objectives:

- To learn basic telescope and binocular terminology
- To understand different types of telescopes and binoculars
- To learn about purchasing a quality telescope or binoculars

Materials: A telescope or binoculars, or a trip to a store that sells them

The history of mankind's understanding of the universe may be broken into two distinct periods: B.T. (Before the Telescope) and A.T. (After the Telescope). No one knows exactly who made the first telescope. Many authorities say it was Jan Lippershey, an eyeglass maker in the Netherlands, who accidentally stumbled upon the idea around 1608. It is certain that Galileo Galilei, an Italian physicist and astronomer, did not invent the telescope, although he is credited with being the first to document its usefulness for studying the universe. With his first telescope in 1609, Galileo discovered craters on the moon and four satellites orbiting Jupiter. He discovered that the planet Venus goes through phases just like our moon. The age of a new universe had dawned.

Since the time of Lippershey and Galileo, the telescope has gone through some powerful changes. Countless improvements and many different designs have evolved in the ensuing 400 years. Though all telescopes do not look the same, they all *do* the same thing. They make distant scenes look closer by passing their images through a series of lenses or mirrors. In doing so, the image is brought to a focus (the point where all of the light rays seem to converge) and, in the process, enlarged.

How this is done is where the variation in design comes in. Today's astronomers use one of three basic telescope designs: the **refractor**, the **reflector**, and the **catadioptric**, all shown in Figure 45-1.

Lippershey's and Galileo's first telescopes were refractors, recognizable by their long, thin tubes. A large lens (the objective) is found in the front end of the telescope tube, while the observer looks through a small lens (the eyepiece) at the back end of the tube.

A reflector usually has a larger diameter than a refractor. Instead of a lens in the front of the tube, a reflector has a large mirror (the primary) buried deep down toward the back of the tube. Though it may look flat, the primary is actually curved ever-so-slightly inward (concave), like a bowl. The most popular type of reflecting telescope was invented by Isaac Newton, and so is called a Newtonian reflector. In a Newtonian, light strikes the primary mirror and reflects to a small, flat mirror (the diagonal) tilted at 45°. The light then bounces off the secondary at a 90° angle, out through a hole in the side of the front part of the tube, and into an eyepiece.

The third type of telescope is a combination of a refractor and a reflector called a catadioptric. Light passes through a large front lens (sometimes called a corrector plate) and on toward a large primary mirror at the back of the tube. Bouncing off the primary, the light reflects toward the front of the tube, where a secondary mirror returns it toward the primary. A small hole in the middle of the primary lets the light pass through and out the back of the telescope and into an eyepiece.

All telescopes regardless of type share many common functions and terminology. For instance, a telescope's size is always referred to by its **aperture**. This is simply the diameter (usually expressed in inches, centimeters, or millimeters) of the instrument's main optic. A 3-inch (80-millimeter, or 80mm) refractor has an objective (or front) lens 3 inches across. The mirror in a 4-inch (100mm) reflector measures 4 inches in diameter. (To convert back and forth between inches and millimeters, remember that there are 25.4 millimeters in each inch. Likewise, to change between centimeters and inches, there are 2.54 centimeters to an inch.)

The **power** or **magnification** of a telescope will change depending on what eyepiece is being used. Figuring out a telescope's magnification or power is easy. To do that, we must know two important numbers: the telescope's focal length (usually stated right on the telescope or in its instructions) and the focal length of the eyepiece (usually stamped on the eyepiece itself). Different eyepieces have different focal lengths, but all are usually specified in millimeters. Look on the barrel of an eyepiece.

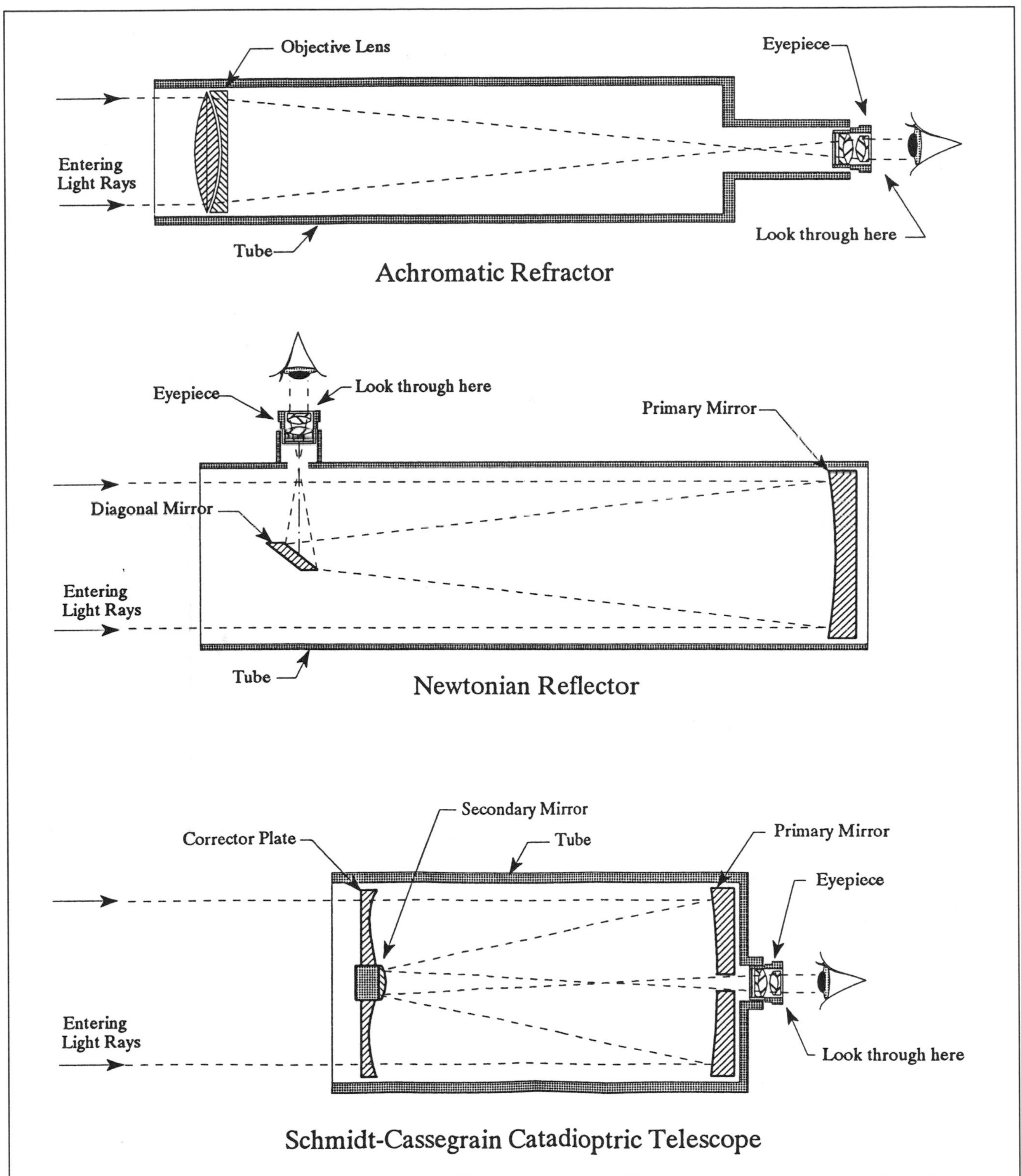

FIGURE 45-1. Cross-sectional views of examples of the three basic types of telescopes (from top): refractor, reflector, and catadioptric.

You might see a number like "K25mm" or "O12mm." Don't worry about the letter (in these cases, K and O); they specify the type or design of eyepiece. The important numbers for now are 25mm and 12mm, signifying 25-millimeter and 12-millimeter focal lengths, respectively.

Using the principles and definitions found in the last activity, let's look at an example of a real telescope, perhaps like a telescope you or a friend may own. An 80-millimeter (3-inch) refractor with a 960-millimeter (38-inch) focal length has a focal ratio of f/12 (since 960 divided by 80 equals 12, as does 38 divided by 3). Forget about the fractions in this case; all the numbers above have been rounded off. Therefore, this telescope might be specified as an 80-millimeter (or 3-inch) f/12 refractor.

Most toy stores and department stores do NOT sell high-quality telescopes. Many camera stores sell a few top-notch telescopes (such as those seen in Figure 45-2) intermingled with an abundance of poor imitations, making it difficult to tell one from another. We have listed the addresses of some of the more reputable telescope makers in Appendix 2.

Let's figure out the magnification of each of the eyepieces mentioned above if they were used in a 3-inch f/12 refractor. First, be sure that all the units of measure are the same. Both the telescope and eyepiece focal lengths must be in either inches, centimeters, or millimeters; the units cannot be mixed. In this case, we know that the telescope's focal length is 38 inches, which is the same as 960 millimeters (remember, 38 x 25.4 = 960). To figure out the magnification of an eyepiece in this telescope, divide the telescope's focal length by the eyepiece's focal length. Therefore the 25mm eyepiece would supply 38X (38-power), while the 12mm eyepiece would supply 80X (80-power).

The remainder of this section is really not a "get-up-and-do-something" activity but more of a "food-for-thought" discussion about choosing a telescope. Many people are under the mistaken impression that the higher the magnification, the better the telescope. This probably goes back to the "bigger is better" philosophy used in just about all aspects of our lives. But too much magnification can actually be bad. You may have been in a store and seen a telescope packaged as a 400X60mm telescope. Translated into English, this means the telescope's aperture is 60 millimeters (2.4 inches), with a magnification of 400X. "Wow," you think to yourself, "this must be a great telescope." WRONG!

There is a limit to the amount of *useful* magnification that can be handled by a telescope. The rule of thumb for small telescopes (that is, any telescope with an aperture less than 8 inches) is that magnification should not exceed 50X per inch of aperture. Exceeding that limit will cause the focused image to lose nearly all of its clarity. In fact, unless you are looking at a bright sky object, such as the moon or a planet, use an eyepiece that produces a magnification even lower than that. Continuing with the example of a 3-inch f/12 refractor, the maximum usable magnification in that telescope is 150X (which translates to a 7-millimeter eyepiece).

One of the most important considerations is how faint a star a telescope will detect. This depends primarily on the telescope's ability to gather light. Quite simply, the larger the aperture, the more light gathered. Therefore, if you are choosing a telescope, in general it is best to buy as large a telescope (by large, we mean diameter or aperture, not overall length) as you can afford.

When children first show a budding interest in astronomy, many parents rush into the local toy store to buy an inexpensive telescope ("inexpensive" is a relative term here, as toy-store telescopes can run over $200). They bring the telescope home, help set it up, point it at the moon or a bright star, and look. The image is usually fuzzy at best, and to make mat-

FIGURE 45-2. A family portrait of three of today's most popular telescopes. Left foreground is a 4-inch refractor, left background is an 8-inch catadioptric (Schmidt-Cassegrain), while on the right is a 6-inch Newtonian reflector. (Photo courtesy Celestron International)

ters worse, the tripod wobbles and shakes. It doesn't take too many nights like this for the child to become discouraged, and the telescope is relegated to a corner next to other discarded toys.

The best way to select a good telescope is to seek out the help of a local amateur astronomer. Many communities have astronomy clubs with members who are more than willing to share their expertise. To find out if there is a club near you, contact any nearby planetariums, museums, or nature centers. If this bears no fruit, go to your local library and look at the most recent May issue of *Astronomy* magazine or September issue of *Sky & Telescope* magazine. Each features a directory of North American astronomy clubs.

No one kind of telescope is best for looking at everything. For instance, refractors are best for looking at the moon and planets. But there are refractors—and then there are refractors. The least expensive, called **simple refractors**, feature a one-piece glass (or in some cases, plastic) objective lens. These consistently produce poor-quality images marred by a wide range of optical faults, or aberrations, and should be avoided.

Better are **achromatic** refractors. Achromatic refractors feature a two-piece (correctly termed two-element) objective lens that greatly improves the view. Most refractors sold today are achromatics. But this quality does not come cheap. Inch for inch and dollar for dollar, refractors are more expensive than just about any kind of telescope. Modern day "amateur" refractors typically produce images that are far superior to many older observatory-class refractors. The smallest-aperture achromatic refractor we recommend is a 3-inch, as anything smaller will quickly prove too limiting. Expect to pay over $500 for a good one.

The very finest refractors are **apochromatics** and produce even finer views than achromatics. These are among the most expensive telescopes today, with price tags typically in the thousands of dollars.

Catadioptric telescopes are known for their portability but are also quite expensive. Most of these super-sophisticated telescopes come with a padded footlocker that allows transport from your urban or suburban home out into the country. Many advanced amateur astronomers choose these telescopes because of their portability and because of the wide range of accessories that are available. But these are not beginners' telescopes, as they can also cost thousands of dollars.

Binoculars are always specified by two numbers separated by an "x," such as 6x30, 7x35, 7x50, and 10x50 (spoken as 6-by-30, 7-by-35, 7-by-50, and 10-by-50). The first number tells the binoculars' magnification, while the second states the diameter (aperture) of each of the two objective lenses in millimeters. Therefore a pair of 7x50 binoculars will magnify the image seven times (seven-power), with each of the two objective (front) lenses measuring 50 millimeters (about 2 inches) across.

For those who want to get the most "bang for their buck," a Newtonian reflector is best by far. Its simplicity of design and execution means, when comparing dollars per inch of aperture, you can buy a larger Newtonian reflector than any other type of telescope. The large apertures and relatively short focal lengths of most reflectors mean bright, crisp images of faint sky objects as well as good views of members of the solar system. Still, a good-quality reflector does not come all that cheaply. Outfitted with a couple of eyepieces and a side-mounted finderscope for aiming the telescope, a 6-inch reflector may cost between $400 and $700. Write or call the telescope manufacturers listed in Appendix 2 and ask for their latest product literature before you buy anything.

If all of these costs have made you dizzy, don't worry—you are not alone. Happily there are a couple of alternatives. Rather than spending between $150 and $200 for a department-store telescope that will probably disappoint you, we recommend purchasing a good pair of binoculars for about $100. Binoculars work on the same principles as telescopes. In fact, binoculars may be likened to two refracting telescopes strapped together. Like a refractor, light enters the binoculars' front (objective) lenses and exits through a pair of eyepieces. In between, it bounces through a series of prisms that flip the image around so that when it leaves the binoculars, everything is right-side up (astronomical telescopes turn everything upside down).

The best binoculars will come with **fully coated** optics to reduce lens flare, prisms made from Bak-4 glass to produce brighter images, and a built-in tripod socket or bracket for attaching to a camera tripod (strongly recommended if more than one person will be using the binoculars at the same time). For a first pair, either 7X35 or 7X50 binoculars are probably best. Avoid zoom binoculars, permanent-focus binoculars, and all models that feature a "quick-focus" seesaw-like focusing lever.

Another alternative to purchasing a high-priced telescope is to build your own. Surprisingly, it's not as difficult as it might sound at first and actually makes a great weekend project for parents and children. Either a refracting or a reflecting telescope can be made from commonly available optics and some spare parts you might have at home right now. The following activity tells you everything you need to know to build a Newtonian reflecting telescope. Why not give it a try?

46 Building Your Own Reflecting Telescope

Level: Advanced

Objective: To construct a 6-inch Newtonian reflecting telescope

Materials:
- Cardboard tube, 8-inch inside diameter by 4 feet long
- Piece of 3/4-inch exterior-grade plywood, 4 feet square
- 3/8-inch by 2 1/2-inch carriage bolt with wing nut and 2 washers
- Two 6-inch by 1-inch PVC rings
- Seven quarter-inch-thick pieces of Teflon, 1-inch square

Tools:
- Handsaw, screwdriver, hammer, jigsaw, power drill, 1/8-inch and 3/8-inch drill bits, 5/8-inch countersink drill bit, 1 1/2-inch hole saw
- Sandpaper
- Carpenter's glue
- Contact cement
- Piece of Formica, 6 inches square (slightly rough surface finish preferred)
- #8 x 2-inch wood screws, one box
- #6 x 1/2-inch flathead wood screws, one box
- Small tube of silicone adhesive
- 1-inch finishing (headless) nails
- Wood sealer (polyurethane varnish)

Paint:
- White primer, flat black, enamel (your choice of color)
- Paint thinner
- Wood filler
- Paint brush (2-inch)
- Dust mask, gloves, and smock

Mail-Order Parts List **(see Appendix 2 for suppliers' addresses):**
- 6-inch f/8 primary mirror (Orion Telescope Center #4500)
- Primary mirror mount (Novak Series "A" mirror mount)
- Secondary mirror (Orion Telescope Center #4602)
- Eyepiece focusing mount (Orion Telescope Center #13014)
- Secondary mirror holder (Orion Telescope Center #4706)
- 6 x 30 finderscope (Orion Telescope Center #13023)

Eyepieces:
- 28mm focal length (Edmund Scientific #P30,787)
- 12mm focal length (Edmund Scientific #P30,940)

BACKGROUND

Observing the moon, planets, and other celestial objects for the first time through a telescope can leave a lasting impression. But viewing such jewels through a telescope you made yourself can leave one exhilarated and proud, ready to dig deeper into the heavens.

FIGURE 46-1. A home-built Dobsonian reflecting telescope will reward you with countless impressive views of the heavens.

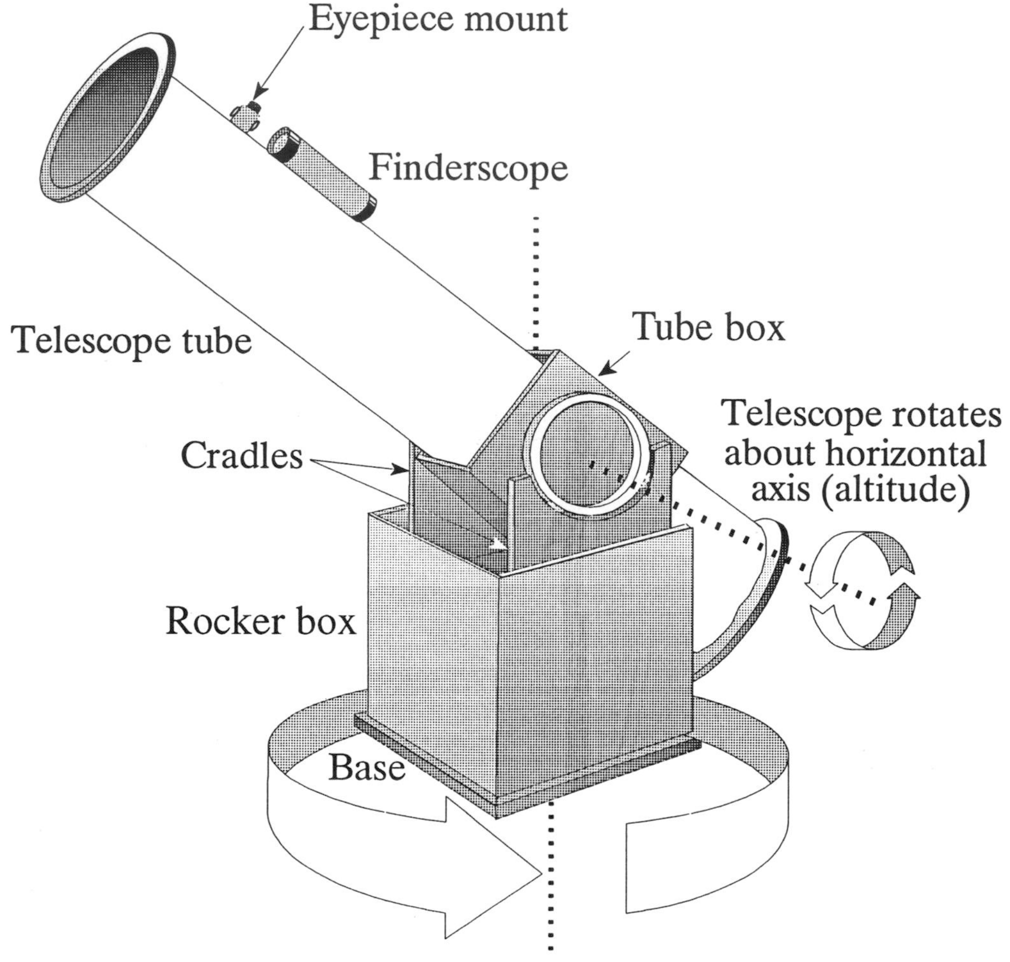

This activity gives plans for constructing a 6-inch Newtonian reflecting telescope (see Figure 46-1). The telescope is supported on a Dobsonian-style mounting. Dobsonian mounts, named after their designer, Californian amateur astronomer John Dobson, are very popular among amateur telescope makers. The general design and layout of our 6-inch telescope are modeled after a larger instrument suggested by Richard Berry in his book *Build Your Own Telescope* (Scribner's, 1985). You will be amazed at the views visible through such an instrument.

ACTIVITY

- Before starting construction, place all the orders for the mail-order items noted in the materials list. We have suggested some sources for each, but feel free to substitute as you see fit.
- Purchase a 4-foot length of 8-inch inside diameter (I.D.) cardboard tube from a local home-supply outlet, lumber yard, or hardware store. The more commonly available brands of tubes are Sonotube and Permatube. You can probably buy a 4-foot length for $10 or less.
- These tubes come with a waxy finish that must be removed to allow paint to stick to the surface. We suggest using a solvent or paint thinner, but be sure to wear gloves and work outdoors for extra ventilation. Afterwards, stand the tube on end and let it dry thoroughly.

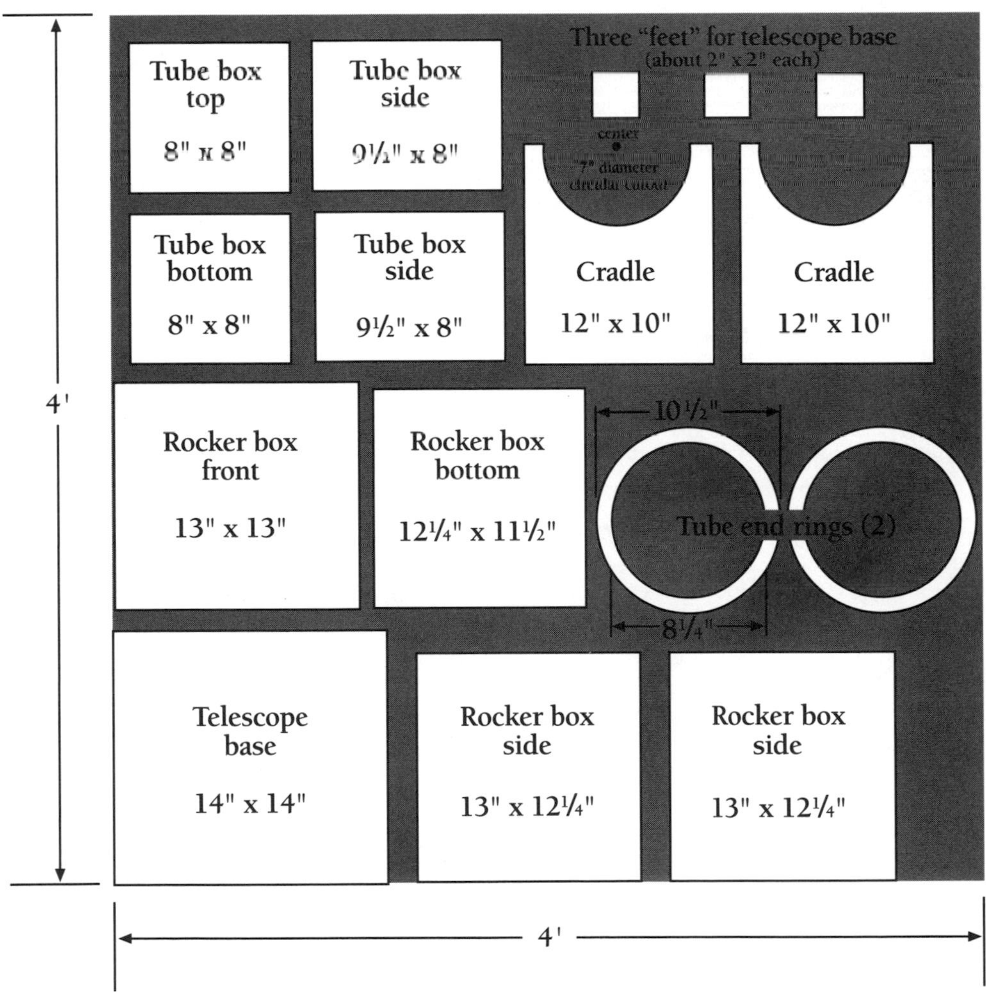

FIGURE 46-2. **All the wood needed for a 6-inch Dobsonian reflector can be cut as shown from a 4' x 4' x ¾" sheet of exterior grade plywood (note the cutouts for the end rings and cradles).**

- While the tube is drying, lay out all of the patterns shown in Figure 46-2 on a 4-foot-square sheet of ¾-inch plywood. For now, just cut out the two tube end rings; the rest can be done later. Take extra care when cutting the inside diameter of the rings, as they must slip snugly over the outside of the tube (they prevent tube deformation). Sand the rings until they are smooth.

- It's now time to install the two tube end rings that you have cut out of plywood. Fit one tube end ring over each end. Carefully drill 1/8-inch holes through the rings and into the tube. Glue and screw the rings to the ends of the tube using #6 x 1/2-inch flathead wood screws. Put wood filler over the screw heads and sand smooth.

- Carefully locate and cut a hole in the side of the telescope tube for the eyepiece mount using a 1 1/2-inch hole saw. Position the hole about 3 inches in from the front of the tube, as indicated in Figure 46-3. For a focused image to form at the eyepiece, the distance from your eye to the surface of the primary mirror must equal the mirror's focal length (here, 48 inches), but we'll worry about that later.

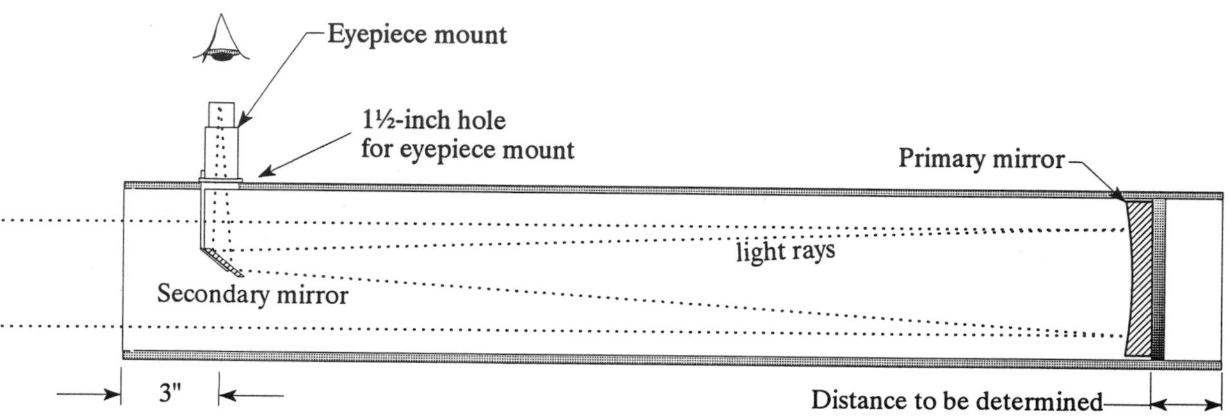

FIGURE 46-3. The light path within the telescope and the placement of its mirrors and eyepiece mount.

- The finderscope helps the astronomer aim the telescope toward the target, much like a sight on a gun. To attach the finderscope to the telescope, simply position its mounting rings where you want the finder to be and drill corresponding screw holes. Drill the holes now, but do not put the finder's mounting rings on yet.

- To strengthen the tube and keep moisture from damaging the cardboard, brush on two coats of polyurethane wood sealer over both the inside and outside of the tube. Be especially liberal when coating the tube ends and the inside of the eyepiece hole. Sand the outside of the tube lightly between coats to get a smooth finish. (For the inside of the tube, tape a paint brush to the end of a broom handle.)

- Cover the tube and rings in and out with a coat of white primer. Do this by standing the tube on one end and painting only the upper half of the tube. Give the paint at least a day to dry, then flip the tube over and paint the other half.

- After the primer has dried, paint the inside of the tube with two coats of flat black paint. Paint the outside of the tube with two coats of enamel, your choice of color. As before, paint the tube in segments, standing it on end and letting it dry each time.

- Begin working on the mounting by cutting out the remaining plywood pieces for the mounting according to the dimensions shown in Figure 46-2. First, make the rocker box. Cut out its four sides from the plywood as shown in the drawing. Glue a 12 1/4-by-13-inch piece of Formica to the bottom of the rocker box using contact cement. When the cement is fully cured, drill a 3/8-inch-diameter hole straight through the center of the Formica and the rocker box bottom.

- Center and countersink four holes in both 12 1/4-inch lengths of each rocker box side as shown in Figure 46-4. Holding the base in a vise, line up each 12 1/4-inch box side with the matching 12 1/4-inch length on the base. Drill and countersink where shown on the drawing using a 1/8-inch drill bit.

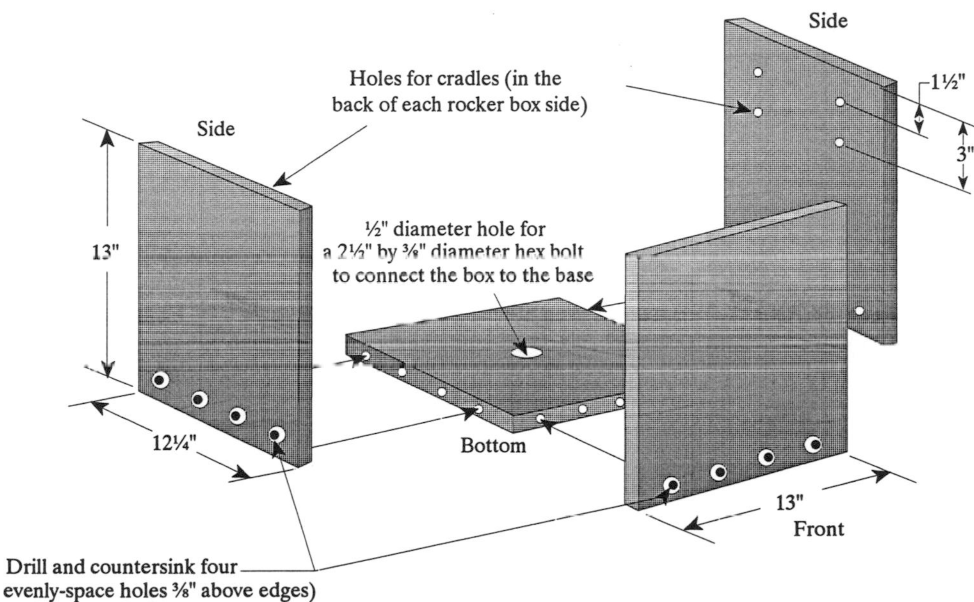

FIGURE 46-4. Assembly of the rocker box. For carrying, handles may be attached to both sides.

- On the rocker box sides without countersinks, center each cradle lengthwise and mark holes as shown in Figure 46-4. Make sure they are both placed identically, or the tube will not swing correctly in the rocker box. Countersink and drill the four holes in each rocker box side, but do not attach the cradles yet. Assemble the rocker box by screwing the sides and front panels to the base.

- Center and drill a 3/8-inch-diameter hole for the 3/8-inch carriage bolt in the center of the telescope base. Nail and glue the three feet at 120° angles on the bottom of the base (see Figure 46-5). After the glue has set, attach the cradles to the rocker box with screws only (in case they need to be detached and adjusted). Leave these screws uncovered.

- Buy two 1-inch pieces of 6-inch diameter PVC pipe with a 1/2-inch to 1-inch wall thickness. These are available from many plumbing-supply stores. (An alternative is to use a pair of 5-inch PVC "toilet flanges.") Start constructing the tube box by exactly centering the PVC rings on the tube box sides (see Figure 46-6). Clamp each in place and drill half a dozen equally spaced 1/8-inch-diameter holes through the sides of each ring and partially into the tube box. Secure each ring onto its tube box side with long wood screws.

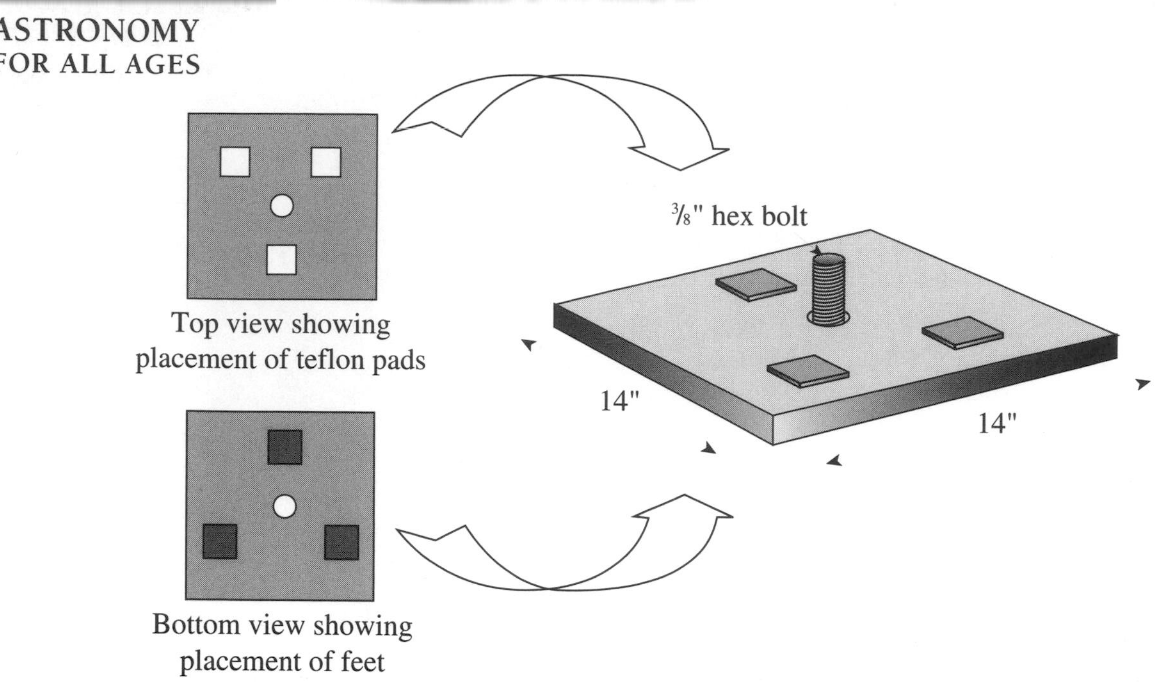

FIGURE 46-5. Feet are added to the telescope base bottom, and small squares of Teflon are nailed to the top (be sure to tap the nail heads below the Teflon top). The carriage bolt goes through the bottom of the rocker box to adjust the tension (like a lazy-susan).

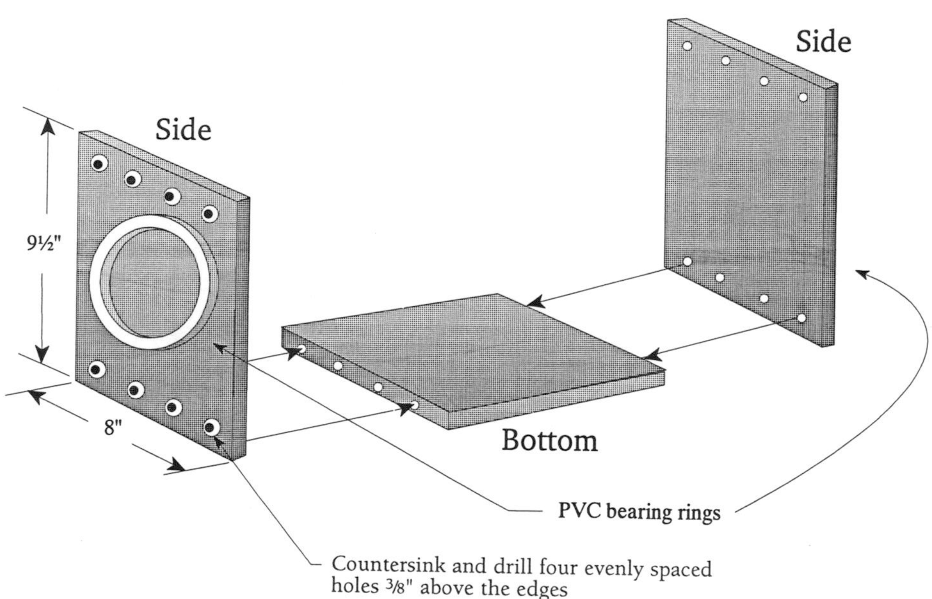

FIGURE 46-6. Assembly of the tube box. Carefully align the PVC rings at the sides' exact center. Then drill small holes through them to the wood, and nail and glue them in place. Leave the box top off until the tube is completed.

- Assemble the tube box by clamping adjacent sides together and drilling ⅛-inch-diameter holes as shown in Figure 46-5. When done, glue and screw only the sides of the tube box to its bottom panel. Leave the top off for now. Eventually the telescope tube will be put into the tube box and the top will be screwed into place. Cover the screw heads with wood filler and let dry, then lightly sand smooth. Give the telescope base, rocker box with cradles, and tube box and its top (still unattached) two coats of sealer, sanding lightly between coats. Apply a coat of durable enamel to these items and set them aside before handling.

- Using small finishing nails, attach three equally spaced Teflon pads to the top of the telescope base. Fit the carriage bolt with washer up through the underside of the base. Line up the center hole in the rocker box with the carriage bolt, sliding the base's bolt through the rocker box's hole. Place the second washer and wing nut over the bolt end and tighten. Don't tighten the wing nut too much, as it will make the instrument difficult to turn. Nail two other pieces of Teflon to each curve along the cradles.
- It's time to put the telescope's optical tube assembly together. As you do the next few steps, be extra careful *not* to touch the mirror's reflective surfaces. Carefully put a drop of silicone adhesive on the back of the small secondary (elliptical) mirror. Place the secondary, reflective side up, on the single metal vane's rectangular plate, as shown in Figure 46-7. Make sure that the mirror's long dimension (the major axis) lines up with the metal vane. Prop the mirror and vane up against something and let the adhesive dry

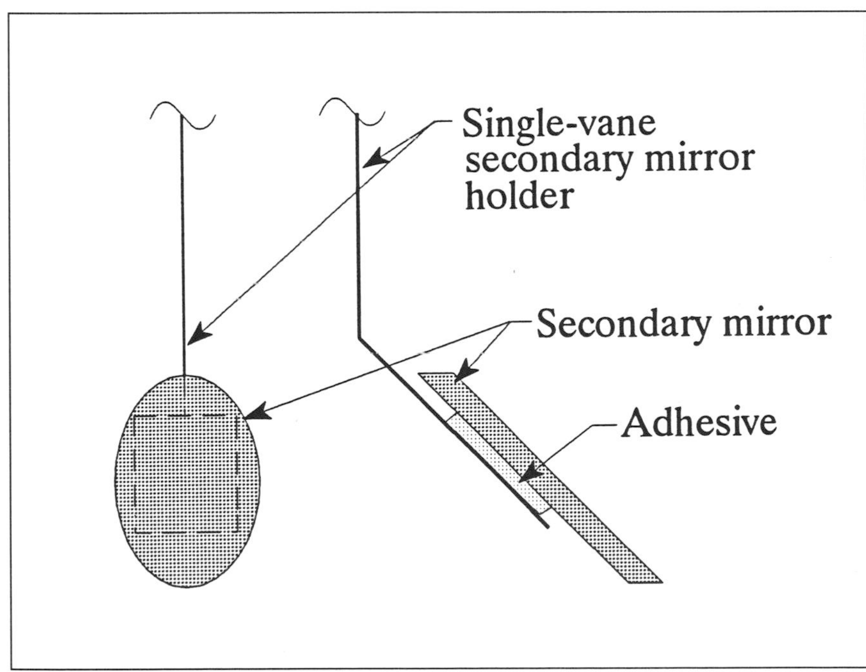

FIGURE 46-7. Glue the diagonal mirror to its holder as shown.

- Afterwards, slide the vane into the small hole on the side of the focusing mount and screw both onto the telescope tube over the 1½-inch hole. Move the secondary mirror on its vane up and down until it is located at the exact center of the tube. Look through the eyepiece holder, and turn the secondary left or right until you can see the bottom end of the tube in its reflection.
- Carefully place the primary mirror into its mount according to the mount's instructions. Now comes the tricky part. In order for the telescope to show everything clearly, the

primary mirror must be positioned exactly. If the mirror is too far forward or too far back in the tube, images will not appear focused. Here's how to work this out.

- Begin by setting up the telescope on its mounting outside on a clear day. Put the topless tube box in the cradle, making sure the PVC rings rest on the Teflon and the telescope box does not scrape against the cradle sides. Place the tube in the open box about two-thirds of the way in from the tube's front end. The fit should be snug enough to keep the tube from slipping.
- Have an assistant hold the primary mirror mount about 4 inches inside the back end of the tube with the mirror facing front. Make sure your helper doesn't accidentally touch the mirror! Next, turn the eyepiece mount's tube in about halfway. Place the 28mm (low power) eyepiece in the mount and aim the telescope toward a distant terrestrial sight such as a treetop or far-off building. If the image is not in focus, have your assistant move the primary mount forward or back in the tube. When the sharpest image is seen in this eyepiece, switch to the 12mm (high power) eyepiece and repeat the process. If the image is fuzzy, try focusing it by turning the eyepiece tube in and out. When both eyepieces show clear images, mark the position of the primary mirror mount in the tube with a pencil. Remove the primary mirror and drill holes in the back of the tube to secure the mirror mount. Make sure that the mirror mount is square to the tube.
- Place the primary mirror and its mount back into the tube, this time securing them with screws, washers, and nuts. Put the finderscope in its mounting rings and place one of the eyepieces in the eyepiece holder. To balance the telescope tube in the tube box, move the tube forward or backward in the box until it doesn't tilt one way or the other. Once the balance point is found, attach the telescope box top with screws.
- To get the best performance from your new optical instrument, the optics must be in line, or collimated, with each other. We've already collimated the secondary mirror by moving it back and forth, left and right in the focusing mount until it reflected the entire back end of the tube. Double check to see that it hasn't moved. For the primary mirror's alignment, remove the eyepiece and look into the eyepiece holder. If everything is lined up correctly, the image of the secondary mirror should be visible in the center of the primary, as in Figure 46-8. If not, carefully turn the adjustment screws in the back of the primary mirror's mount, following the manufacturer's instructions. Continue to alternately tighten and loosen the adjustment screws until the primary and secondary are collimated. It is critical that you do not try to rush this adjustment. A properly collimated telescope can make all the difference in the world in how well the telescope will perform—or even if it will work at all!
- When all is in order, take your new telescope outdoors for its christening. Leave it out at least an hour before you plan to use it so that any built-up heat in the tube escapes. This reduces the fuzziness of images due to heat currents in the tube. When not in use, the telescope's optics must be protected from dirt and dust. Seal the tube with shower caps or some type of lid at both ends.

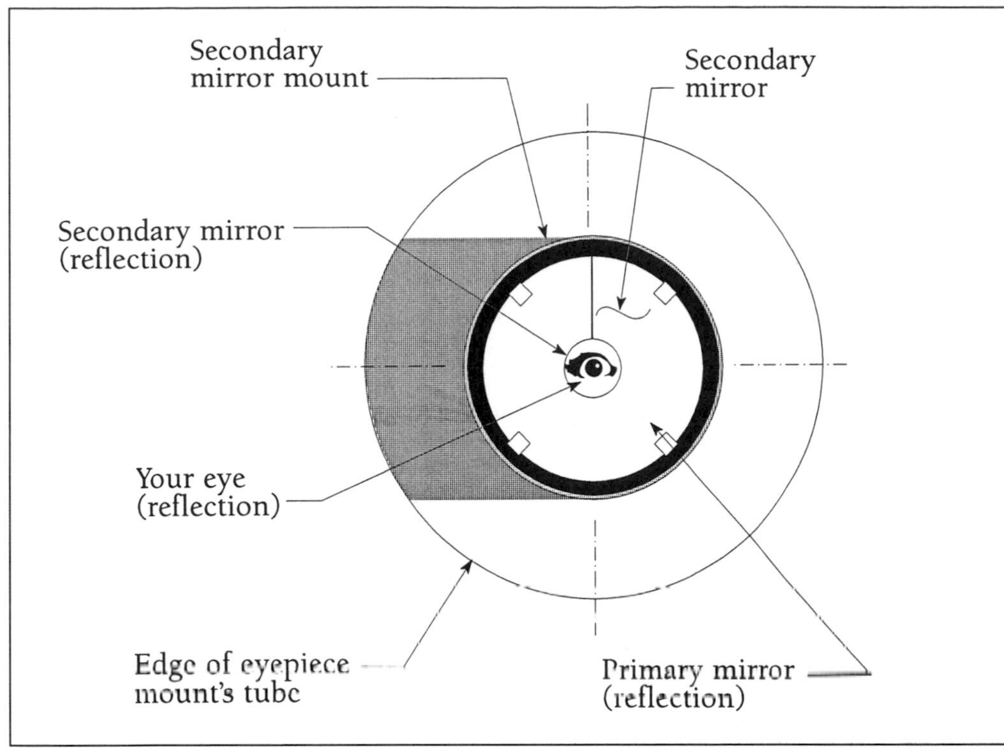

FIGURE 46-8. In order for the telescope to focus correctly, the primary and diagonal mirrors must be aligned (collimated) as shown.

- Seal the open eyepiece holder with an empty plastic can from a roll of 35mm film.

Viewing through a telescope is always a fascinating experience. But using one that you made yourself adds a special magic that can only be appreciated once the telescope is done. Why not give it a try?

47 Discovering ROY G. BIV

Building Your Own Spectroscope

Level: Intermediate and advanced

Objective: To learn about the composition and wave nature of visible light

Materials:

- A shoe box
- Flat black spray paint (or a few sheets of black construction paper and Scotch tape)
- A pair of scissors and a ruler
- 1-inch-wide black electrical tape
- A diffraction grating (see Appendix 2 for suppliers' addresses)

BACKGROUND

Although space travel is easily done in science fiction, astronomers cannot travel to the stars to study their features. Instead they use a wide variety of instruments attached to telescopes to help them investigate these objects by observing the light they emit. In this way, astronomers can tell if a star has an otherwise unseen companion, how fast the star is moving toward or away from Earth, and even what elements make up the star (such as hydrogen, helium, and calcium). In this activity, you will construct a **spectroscope** ("spectro" referring to light and "scope" to the housing for the instrument), a device that astronomers use to break light into its component colors.

A **diffraction grating** will be the centerpiece of your spectroscope. In appearance, it resembles a photographic slide, though with some important differences. You'll notice that the grating is transparent—and that it looks scratched. This is perfectly normal. You are seeing the fine, minute scratches, or grooves, scribed in the grating's acetate or glass. Most commercial gratings—with 13,000 grooves per inch—cost about 50 cents or a dollar. Holographic gratings with nearly 20,000 lines per inch give brighter images and cost about $5.00 apiece.

What does a diffraction grating do? The grating demonstrates a property of light waves that allows them to seem to go around corners , an effect similar to the more familiar bending (**refracting**) of light. To understand this, think of waves of water approaching a shoreline and a pier. As they approach, the waves far from the pier's pilings reach the shore unobstructed. But the waves that hit the pilings are partially blocked and spread (or **diffract**) around the pilings.

Diffraction also occurs with light waves and sound waves. Physics students often encounter an experiment in which they send a laser beam toward a human hair, using the resulting diffraction pattern to measure the hair's thickness. With sound, it is diffraction that allows you to hear music around a corner. In the diffraction grating, unscratched slits between each dark groove serve to separate light into its component colors, the familiar rainbow pattern known as the **visible spectrum**.

ACTIVITY

To build your spectroscope, cut a 1¼-by-1-inch rectangle out of each end of your shoe box to be used as "windows" into which light will pass (Figure 47-1). You now have an opening centered in each end of the box. Next, make the inside of the box and lid as dark as possible, using either cut and taped black construction paper or (more easily) flat black spray paint.

Once the box interior is blackened, create a narrow, vertical slit in one of the "windows" by using pieces of black electrical tape. Do this by sticking one piece of tape partially over the left side of the window and one piece of tape partially over the right side of the window, leaving about a ⅛-inch vertical gap, or slit, between them. (Also stick matching pieces of tape on the interior side of this window to help reinforce the slit.) It is through this slit that light will enter the box, passing to the window at the other end, through which you will be looking.

Now for the finishing touch. Take your diffraction grating and tape it over the fully open window at the other end of the shoe box. It's OK to tape the grating either inside or outside the box and either vertically or horizontally.

The spectroscope will now allow you to examine a light source to reveal the color components of visible light. Hold the covered box with the diffraction grating at your eye. Aim

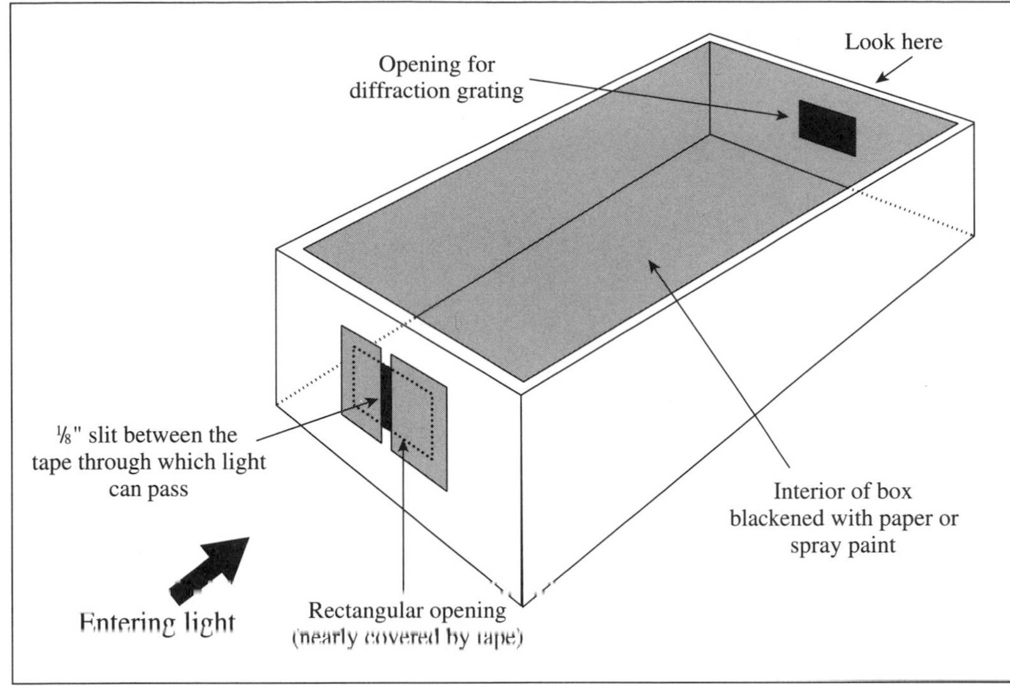

FIGURE 47-1. **The basic design of the shoebox spectroscope (the box top has been removed for clarity).**

the narrow, vertical slit at a light source, such as a lamp bulb, and move the box so that the slit is near the light—**but do not look at the slit**. Instead, look to the left or right, or above or below, the slit. You should now see the visible spectrum ("rainbow") of colors that make up white light.

You can also use the spectroscope to observe the sun, but remember: DO NOT POINT THE SLIT DIRECTLY AT THE SUN! Instead, point the slit toward blue sky *near* the sun, or do this indoors so that a window frame blocks the direct sunlight. Either way, you will still see the sun's spectrum.

As you look at a spectrum in your spectroscope, note the arrangement of colors relative to the slit of light. You should see that red appears farthest from the slit (somewhere near the inside corner of the shoe box), whereas blues and purples are closest to the slit. Because a wave of red light is longer than a blue one, red light is diffracted more than blue. As a rule, longer waves are diffracted more than shorter ones. Also notice the order of the colors, moving inward toward the slit from red. You can easily remember this order by using the first letters of each color to spell out the name ROY G. BIV: Red, Orange, Yellow, Green, Blue, Indigo, Violet.

A similar rainbow effect is visible through a small triangular piece of glass called a prism (Figure 47-2), which illustrates a property called **dispersion** (not diffraction). The visible spectrum can also be seen through a spray of water from a garden hose, with your back to the sun and the shower of water above you. The spray of water illustrates the property of light known as **refraction** (not diffraction) and is one more demonstration of the fact that light is composed of all colors, a proposition set forth by physicist Isaac Newton three centuries ago.

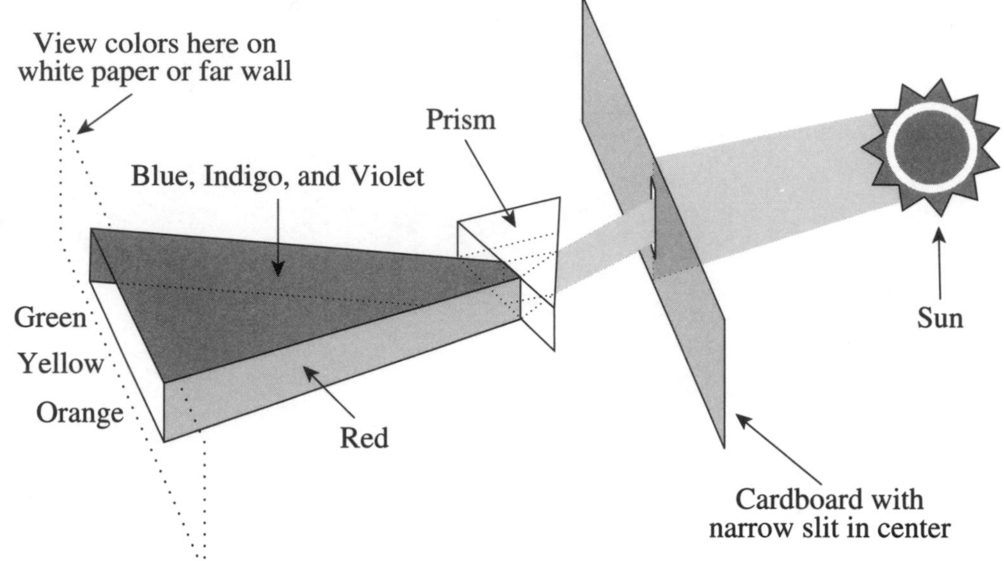

FIGURE 47-2. White light (such as sunlight) entering a prism is spread into its component colors. Called dispersion, blue and violet are dispersed the most, while red is dispersed the least.

48 Understanding Your Camera

The World of Astrophotography

Level: Intermediate and advanced

Objective: To learn about your camera and to use it to photograph the stars

Materials:

- 35mm single-lens reflex (SLR) camera
- A manually focusing lens
- A sturdy tripod
- A cable release
- Slide film, such as Kodachrome 64 or Fujichrome 100
- A clear night with stars or the moon out
- A red-filtered flashlight (see Activity 2)

BACKGROUND

Many of us have experience with a common point-and-shoot camera, though few of us have ever considered using a camera to photograph the night sky. You might think it takes much expensive equipment and hours of photography to capture pictures of the universe.

But taking photos of the constellations, planets, and the moon can be easy and fun. With the proper conditions, photos such as Figure 48-1 are within your reach (take a look also at Figure 20-1 and 20-2 in Activity 20). In this activity you will learn some of the basics of astrophotography, taking pictures of the stars.

FIGURE 48-1. A wide-angle photo of the Summer Milky Way taken by Alan B. Morton atop Mt. Graham, Arizona. The brightest patch lies between the constellations Sagittarius and Scorpius. (Courtesy of Dennis Milon)

First, you will need a single-lens reflex (SLR) camera (Figure 48-2). Such a camera is designed with a viewfinder that allows you to look directly through the camera lens so that you can see precisely what the camera sees when you take a picture. Many photographic

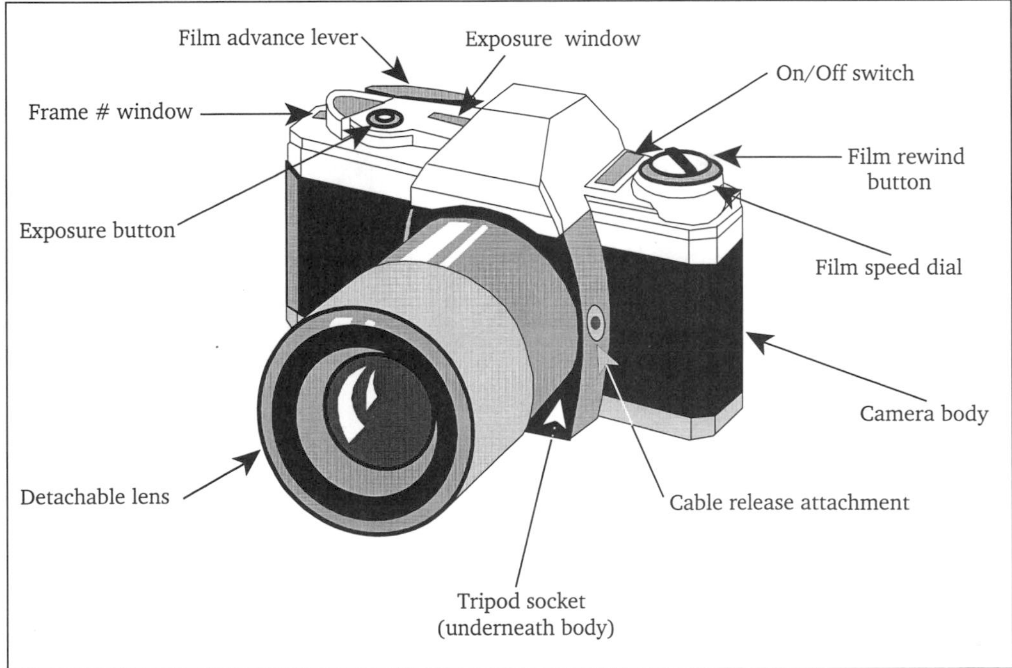

FIGURE 48-2. The basic features of a typical single lens reflex (SLR) camera.

accessories are designed around the basic structure of the SLR camera, making it one of the most popular camera types. Get your camera, and let's take a quick rundown of its features.

Load film into the camera. On the camera, look for the film speed dial. The "speed" of film refers to how sensitive the film is to light. Certain bright-light conditions may require the need for slow film (also called "fine-grained" film, as it looks very clear upon developing), whereas exposures of the dark sky will require "coarse-grained" higher-speed film that is more sensitive to light. For example, a film of speed 64 is called "slow" because it is not very sensitive to light (it requires a lot of light to make a picture), whereas film rated 1600 is very sensitive to light (it doesn't need much light to make a picture). Typical outdoor photos are shot using films of 25 to 400 speed, but in astrophotography, typical film speeds range from 64 all the way up to 3200. Set the film speed dial to match the speed that is listed on the package of film you are using.

Now look on the top of your camera for a round, rough-edged dial that may look similar to that shown in Figure 48-3. Called the exposure dial, this controls the device within the camera (called the shutter) that opens and shuts, permitting light to fall on the film, thus exposing the film. Each number on the dial represents a fraction of one second. For example, "125" is really 1/125 second, "2" is really 1/2 second, and "1" is one second. When this dial is set to 125, the camera shutter will stay open for 1/125th of a second when you snap a picture.

FIGURE 48-3. Each number on the exposure dial shown here controls the amount of time light falls onto the film in the camera.

For most photos you take at night, you will set the exposure dial to the "B" setting. Called the "bulb" setting, this permits you to take as long an exposure as you like. These long exposures are taken not by holding your finger on the exposure button, but by using a cable release (see Figure 48-4). The cable release screws into a threaded hole on the camera body (usually on the camera front or on the exposure dial) and has a plunger at one end. The plunger can be pushed in and locked, allowing the camera shutter to stay open until you release the plunger.

The camera device that focuses light onto the film is the lens. One beauty of modern cameras is that their lenses are detachable, allowing you to change lenses at will in order to obtain a variety of different photos. Look on the front of a lens for a number followed by "mm," such as 35mm, 50mm, or 150mm. This refers to the focal length of the lens, in millimeters. The lower the number (that is, the shorter the focal length), the larger an amount of area your picture encompasses (called field of view). Similarly, the higher the number (that is the longer the focal length), the smaller the field of view. With a long-focal-length lens (a so-called telephoto lens), the subject will appear much closer than with a shorter-focal-length lens—as you can see in the two photos of the moon in Figure 48-5.

On the side of the lens casing, you will see a succession of numbers, including 4, 5.6, 8, and 11. These are known as f-stops (Figure 48-7). You can set whatever f-stop you believe is appropriate for a particular photo. The f-stop indicates the relative amount of light the lens will let in. A stop of f/4 lets in much more light than f/11 or a higher f-stop. The f-stop controls how wide the lens iris (or diaphragm) will open (Figure 48-6). When in a rural, dark location, you will use stops of about f/1.7 to f/4.

FIGURE 48-4. When the camera is attached to a tripod, exposures can be made using a cable release, shown here. Cable releases also permit very long exposures to be taken.

FIGURE 48-5. By using lenses of different (or varying) focal lengths, you may enlarge or reduce the size of your subject, as shown here with the first quarter moon (using Kodachrome 64 film). Left exposure is at f/5.6 for 1/60 second at a focal length of 135 mm, whereas the right exposure is at f/11 for 1/30 second at 1000mm focal length.

FIGURE 48-6. As shown, an f-stop of f/3.5 (left) widens the lens iris to increase the amount of light entering the camera. An f-stop of f/11 (right) has the opposite effect, reducing the amount of light entering the camera.

ACTIVITY

A good subject to begin with is the moon as it is full of detail (best at the quarter or crescent phases), is easily located, and requires only brief exposures. With your camera securely attached to a sturdy tripod and the moon centered in the viewfinder, try taking several shots at a variety of exposures (called **bracketing**). If you like, keep a fixed f-stop and vary the exposure time. To achieve different image sizes without having to buy four or five lenses, you can purchase a zoom lens, which will cover a whole range of focal lengths (such as 28mm–100mm or 80mm–300mm). You can change focal length at the touch of a finger and thus add more creativity to your shots. Try the exposures described in Figure 48-5, create combinations of your own, or see Activity 50 on shooting the moon.

Other impressive and easy-to-shoot astronomical subjects are **planetary** and **lunar conjunctions**, when planets and the moon appear close to one another in the sky (Figure 28-2 on p. 105). Such events are often described well in advance in *Sky & Telescope* magazine,

Astronomy magazine, and the *Sky Calendar* (see Appendix 1). In these situations, a well-composed photograph can be easily obtained if the moon is near the horizon or some other interesting foreground objects. In general, slide films (rather than print films) tend to develop more accurately what was seen by the camera and are thus often the choice for astrophotography.

On a dark night away from city lights, you may be able to get photos of the Milky Way or of individual constellations if you have sufficiently fast film (such as Fujichrome 1600). Since Earth is constantly spinning (and your camera on its tripod is fixed to Earth), stars will appear to cross the sky during the night. As a result, any photos you take will begin to show this movement of the stars after an exposure of about 30 to 45 seconds (for lens focal lengths of about 35mm). This apparent star movement results in **star trails** (see Activity 49). Once you've practiced getting familiar with your camera in the dark, photographing celestial objects will become second nature. When you become interested in attaching your camera to a telescope, consult Activity 50 as well as the references in Appendix 1.

49 Showing That Earth Spins

Making Photographic Star Trails

Level: Intermediate and advanced

Objectives: To discover Earth's rotation, the North Star's stationary position, and star colors

Materials:

- A clear, moonless night far from lights
- A camera with a "B" (bulb) exposure setting, a cable release socket, and a focal length between 28mm and 50mm
- A cable release
- A sturdy tripod
- Ektachrome or Fujichrome slide film, with a speed between 64 and 400
- A red-filtered flashlight (see Activity 2)
- Notepad, pen, and watch
- Activity 48 (for reference)

BACKGROUND

Go outside on a clear night about an hour after sunset and note the positions of particularly bright stars compared to the horizon or nearby treetops. Returning a few hours later, you will notice that the stars appear to have changed position. Most stars will appear to have drifted westward.

This apparent westward motion of the stars is a direct result of Earth's spinning (or rotation) on its axis, called its **daily** or **diurnal** motion. As Earth rotates from west to east, celestial objects appear to move in the *opposite* way, from east to west. To imagine this, pretend you are on a merry-go-round. As it rotates, objects that are *not* on the merry-go-round will appear to move *against* the direction in which you are spinning.

Using a camera whose shutter can be kept open for long periods using a "bulb" setting, you can capture this effect by taking time-exposure photographs of stars.

ACTIVITY

The decision on whether to use a higher-speed film or a lower-speed film will depend primarily on the location from which you will be shooting. In general, the more light pollution your site has, the lower the film-speed number (the least light sensitive film) you may want to use (such as 64 or 100). But if you are observing from a dark rural location, film with a higher number (such as 400 speed) will do better work. In general, slide film gives better results than print film.

Find a location with a clear, unobstructed view of the northern horizon. Load the film into the camera and set your camera's exposure ring to the "B" (bulb) setting. Attach the camera to the tripod via the threaded adapter on the camera's underside. Screw in the cable release to the threaded hole on the camera, making sure its ring is locked to keep the cable release plunger out when the cable release button is pressed. Use a lens with a focal length between 28mm and 50mm. Set the f/stop between f/1.7 and f/4. Now find Polaris, and point the camera toward it, positioning the star near the center of the field of view.

Now you are ready to take your time-exposure photo. There are many variables involved, but here's a typical configuration that you might try: With 400-speed slide film in a dark, rural location, shoot with a 28mm lens at f/2.8 for eight minutes.

With the tripod firmly placed, the camera turned on, the lens cap off, and the lens focused on infinity (∞), hold your hand about an inch in front of the lens for several seconds as you click the cable release to open the shutter. (Holding your hand in front of the lens prevents light from entering the camera while it is subjected to slight vibrations as the cable release opens the shutter.) Then remove your hand—and the lens will be free to collect starlight on the film. Jot down your f/stop, exposure time, frame number, subject, and so forth. During your timed exposure, avoid wind, bumping the camera, stray light, or blinking airplane lights going through your field of view (block the lens with your hand until the plane passes over). After eight minutes (or whatever other exposure time you've decided on), once again block the lens with your hand as you click the cable release to close the shutter.

When your film is developed, you should see something similar to Figure 49-1, called **star trails**. Because the camera and tripod are held to Earth, they are moving with our spinning planet. The stars, however, are not—and thus appear to leave small circular arcs on your film as Earth rotates. Polaris, however, will appear to be stationary because the north pole is aimed almost directly at Polaris and thus the spinning of the planet will not change its apparent location.

You can also photograph star trails near the eastern or western horizon. You'll notice that stars farther from the North Star in the sky have trails that appear to straighten out. To understand why, watch a spinning bicycle wheel and imagine "stars" on the spokes. Stars closer to the center of the wheel cover less distance as they circle than do outermost ones. But the **angular** rate at which they all move is the same, because all are part of the same wheel. In the sky a similar effect occurs, causing stars farther from Polaris to apparently cover larger arcs than those closer in.

FIGURE 49-1. A three-hour-long photo centered on Polaris reveals an impressive sight. (Photo by Richard E. Hill at Kitt Peak National Observatory in Arizona using Ektachrome 400 film at f/5.6; courtesy of Dennis Milon)

Your photos should also show different star colors, a direct indication of their surface temperatures (hotter ones are bluer, cooler ones are redder). To create eye-catching photos, try composing your pictures with trees, mountains, or other landscapes in the foreground. Making star trails is an enjoyable and instructive way to photograph and discover the starry heavens.

50 Shoot the Moon

Level: Intermediate and advanced

Objective: To learn how to photograph the moon with a telescope or telephoto lens

Materials:

- 35mm single-lens reflex (SLR) camera
- Telescope or telephoto lens
- Cable release
- Tripod
- Roll of moderately fast film
- Dark cloth or towel

BACKGROUND

How did you do with the last activity? Were you successful in photographing the stars? We hope so. There is great satisfaction in capturing a bit of the sky on film. Photographing the stars with a camera set on a tripod is the simplest type of astrophotography. Of course, once you have tasted victory, you are bound to yearn for more! For most astrophotographers, that next step is photographing the moon. No other sky object is photographed more than the moon. Its proximity to Earth means that it reveals more detail than any other object in the heavens. Figure 50-1 shows a photograph of the moon taken through an 8-inch reflector.

FIGURE 50-1. The moon one day after first-quarter phase, as photographed with an 8-inch reflecting telescope at f/8, using ASA 100 film and an exposure of 1/60 second.

ACTIVITY

Photographing the moon is really quite simple if you have the right equipment. A review of the materials listing for this activity shows that equipment consists of a 35mm single-lens reflex (SLR) camera, tripod, cable release, a roll of moderately fast film, and a telephoto lens or telescope. If you choose to use a telephoto lens, it should have a focal length of at least 200mm, with 400 mm or more preferred. Even then, the moon's image will appear quite small. To record the lunar seas, craters, and other surface features, a telescope is usually required.

While there are several ways of photographing through a telescope, only the simplest method will be described here. You are invited to browse through Appendix 1 for books on astrophotography.

The system described here is called the **afocal method**. The afocal method can be used with almost all 35mm SLR cameras and telescopes. Just about any type and speed of film can be used, although slower films (film speed of 200 or less) are usually preferred.

Once the film is loaded into the camera, place the camera with its standard lens on a tripod, as shown in Figure 50-2. Set the lens's focus to **infinity (∞)** and its aperture (f-stop) wide open (that is, at the lowest number). Next, aim the telescope toward the moon. If your telescope has a clock drive on it to track the sky, be sure to switch it on.

Looking through the camera's viewfinder, raise and lower the camera on the tripod (as shown in the photo) until the camera is aimed directly into the telescope's eyepiece, with the moon centered in the camera's field of view. This can be more difficult than it sounds, so take your time. Look through the telescope's side-mounted finderscope to keep the moon centered in view.

With the moon in the camera's viewfinder, adjust the telescope's focusing knob in and out until the camera image is at its sharpest.

Select the proper exposure, drape a dark cloth or towel over the eyepiece focuser and camera to block extraneous light, and you are ready to shoot the moon.

But, wait! What exposure should you select? If your camera has a built-in light meter, do *not* rely on its reading. Odds are it will be wrong because the moon is a small, bright object on a large, dark background. This tends to confuse the light-meter reading. The only way to select the right exposure is by trial and error (with a little help, that is).

FIGURE 50-2. Helen Harrington's telescope set up to take photographs of the moon by the afocal method described in this activity.

To "guesstimate" the proper exposure, we must first know the telescope/camera combined f-stop. There is only one way to find that out; it must be calculated. Don't throw up your hands in disgust; the calculation is really very simple. The effective focal ratio (or EFR as it is called) can be calculated using the following formula:

EFR = (camera lens focal length X telescope magnification) ÷ aperture of telescope

Here is an example. Suppose you want to photograph the first-quarter moon with the 6-inch f/8 Newtonian reflector constructed in Activity 46. You select a 25mm eyepiece and a 50mm f/1.8 camera lens. What's the effective focal ratio? First, convert inches to millimeters or vice versa, so that the units of measure are all the same. Remember, there are 25.4 millimeters to an inch. This means that the telescope's 6-inch aperture is equal to 152mm (6 inches X 25.4 millimeters per inch = 152 millimeters).

The formula in Activity 45 told us that a telescope's magnification is equal to its focal length divided by the focal length of the eyepiece. For this example, the results are as follows:

Magnification = (48 X 25.4) ÷ 25 = 48-power

Finally, to figure out the telescope/camera's EFR, plug these values into the formula:

EFR = (50mm X 48-power) ÷ 152mm = f/16

With the EFR known, we can now take an intelligent guess for the exposure. Table 50-1 itemizes some suggested exposures for some popular films and EFR values (f/2 through f/16). If your EFR values don't match the ones in the table, use the closest values. But take heed, we said these exposures are *guesses*. It is always a good idea to take more than one photograph, varying the exposure, just in case something goes wrong along the way (and it always does).

TABLE 50-1

Some Suggested Exposures for Shooting the Moon

EFR	CRESCENT	QUARTER	GIBBOUS	FULL
		FILM SPEED 50-64		
f/2	1/250	1/500	1/1000	n/a
f/2.8	1/125	1/250	1/500	1/1000
f/4	1/60	1/125	1/250	1/500
f/5.6	1/30	1/60	1/125	1/250
f/8	1/15	1/30	1/60	1/125
f/11	1/8	1/15	1/30	1/60
f/16	1/4	1/8	1/15	1/30
		FILM SPEED 80-125		
f/2	1/500	1/1000	n/a	n/a
f/2.8	1/250	1/500	1/1000	n/a
f/4	1/125	1/250	1/500	1/1000
f/5.6	1/60	1/125	1/250	1/500
f/8	1/30	1/60	1/125	1/250
f/11	1/15	1/30	1/60	1/125
f/16	1/8	1/15	1/30	1/60
		FILM SPEED 160-200		
f/2	1/1000	n/a	n/a	n/a
f/2.8	1/500	1/1000	n/a	n/a
f/4	1/250	1/500	1/1000	n/a
f/5.6	1/125	1/250	1/500	1/1000
f/8	1/60	1/125	1/250	1/500
f/11	1/30	1/60	1/125	1/250
f/16	1/15	1/30	1/60	1/125

Note: In this table, "n/a" indicates that a photograph would likely be overexposed.

Always write down the particulars about each photo you take so that you can learn from your successes and your mistakes. Include such items as date, subject, equipment (camera, plus telephoto lens or telescope), film type and speed, frame number, exposure, f-stop, and sky conditions.

Photographing the moon can be great fun. Begin a collection of moon pictures at each different phase. Compare the appearance of each. Try to identify as many of the craters as possible from Activity 18. How many can you find? Then take your photograph outside and compare it to the view through your telescope.

51 An Astronomical Scavenger Hunt

Level: All

Objective: To discover how astronomy has influenced our culture

Materials: The world around us

BACKGROUND

You are a star! So are we. In fact, so is everyone and everything that ever was or ever will be on this planet. Our bodies are composed of elements that were synthesized inside some distant star long, long ago. Over eons, that star evolved until it used up all of its hydrogen fuel. When that happened, the star became unstable and exploded, hurling its component gases in all directions. These gases slowly combined with other interstellar gas clouds (called nebulae) to form new stars. Our sun was born, along with Earth and other members of the solar system, inside one of these nebulae about 4.5 billion years ago. From the combination of elements, such as carbon, hydrogen, oxygen, and nitrogen, that came from those distant, ancient stars, life on Earth was born and began to evolve. *That's* why we are all stars!

But when we refer to someone as a "star" today, we are probably only talking about a celebrity of some sort. Possibly the star we are thinking of is an actor or actress, a musician, or an athlete. Applying the word "star" to that person is just one way astronomy has influenced our culture. Can you think of others?

This is the basis of a game we call the Astronomical Scavenger Hunt. The object of the game is to come up with as many words, things, or places that are named for the sun, moon, any of the planets, stars, or just about anything else astronomical. You might be surprised to find out just how many things in our daily lives have their origins in the stars.

Speaking of days, take a look at a calendar. The calendar we use today (called the **Gregorian calendar** because its use was first authorized by Pope Gregory XIII in 1582) was formed from the solar system. Back when the calendar was first created, the solar system was thought to be made up of seven objects: the sun, moon, Mercury, Venus, Mars, Jupiter, and Saturn (notice how the Earth was curiously omitted). These seven objects gave us the seven days of the week. **Sunday** is named for the sun, **Monday** comes from the moon, and **Saturday** for Saturn. The other days of the week come from Germanic forms of the planet's names: **Tuesday** from Mars, **Wednesday** from Mercury, **Thursday** from Jupiter, and **Friday** from Venus.

Each one of those days is measured from start to finish by how long it takes the Earth to spin once on its axis (nearly twenty-four hours). Combine about twenty-nine of these days together to get a month. The month was first measured by the orbit of the moon around the Earth, which takes about 29.5 days. Both the word "month" (originally, "**moon**-th") as well as its length has been modified since the calendar was first adopted. Most of today's "moonths" contain either 30 or 31 days to make up for time lost in the original scheme.

Aside from the calendar, astronomy also has greatly influenced geography. There are several places on Earth's surface that were pinpointed by events in the sky. For instance, the north and south poles mark the points of the Earth's axis. The equator is the halfway point between the poles, but also where the sun appears directly overhead at noon on the first day of spring (called the vernal equinox) and of autumn (the autumnal equinox). The Tropic of Cancer marks where the sun lies directly overhead at noon on the first day of summer, while the Tropic of Capricorn marks where the sun is directly overhead at noon on the first day of winter.

ACTIVITY

Take a look around. Many everyday products are named for heavenly bodies found somewhere in the universe. Companies know that astronomy sells! Figure 51-1 shows just a few examples.

FIGURE 51-1. A veritable smorgasbord of astronomical treats. (Photo by Jack Megas)

To help you begin your own search, think of everyday categories of things that might include astronomical terms. Some suggested categories are automobile manufacturers and models, movies, candy, household products, flags of countries and states, sports teams, and songs. Come up with as many names as you can of items in these categories that use celestial words.

Some possible answers appear in Appendix 5—but come up with your own first. No fair peeking! When you are done, compare your results. Who got more items, you or us? If *you* did, congratulations. Give yourself a gold star!

Appendix 1: Bibliography

BOOKS

Apfel, N. *Astronomy Projects for Young Scientists.* Prentice Hall, 1984.

Berry, R. *Build Your Own Telescope.* Charles Scribner's Sons, 1985.

Brown, S. *All About Telescopes.* Edmund Scientific, 1975.

Chartrand III, M. *Skyguide: A Field Guide for Amateur Astronomers.* Western Publishing, 1982.

Cherrington, E. *Exploring the Moon through Binoculars and Small Telescopes.* Dover, 1983.

Covington, M. *Astrophotography for the Amateur.* Cambridge University Press, 1991.

Dickinson, T., Costanzo, V., and Chaple, G. *Edmund Mag 6 Star Atlas.* Edmund Scientific, 1982.

Dickinson, T. *Nightwatch: An Equinox Guide to Viewing the Universe.* Camden House, 1983.

Eastman Kodak. *Astrophotography Basics* (Publication P150). Kodak, 1988.

Espenak, F. *Fifty Year Canon of Lunar Eclipses: 1986–2035.* Sky Publishing, 1988.

Espenak, F. *Fifty Year Canon of Solar Eclipses: 1986–2035.* Sky Publishing, 1989.

Harrington, P. *Star Ware.* John Wiley and Sons, 1994.

Harrington, P. *Touring the Universe Through Binoculars.* John Wiley and Sons, 1990.

Hedgecoe, J. *The Photographer's Handbook.* Alfred A. Knopf, 1982.

Kaufmann III, W. *Universe.* W.H. Freeman and Co., 1986.

Kitt, M. *The Moon: An Observing Guide for Backyard Telescopes.* Kalmbach Books, 1992.

Littman, M., and Willcox, K. *Totality: Eclipses of the Sun.* University of Hawaii, 1991.

Littman, M. *Footsteps: In Honor of the Apollo Moonflights.* Hansen Planetarium, 1979.

Mayall, N., et al. *Sky Observer's Guide.* Western Publishing, 1985.

McAleer, N. *The Cosmic Mind-Boggling Book.* Warner Books, 1982.

Menzel, D., and Pasachoff, J. *Peterson's Field Guide to the Stars and Planets.* Houghton Mifflin, 1993.

Norton, A. *Norton's 2000.0 Star Atlas and Reference Handbook.* Longman Scientific & Technical, 1989.

Ottewell, G. *Astronomical Calendar* (annual). Greenville, S.C.: Furman University.

Pasachoff, J. *Astronomy: From the Earth to the Universe.* Saunders College Publishing, 1979.

Percy, R., et al. *Observer's Handbook* (annual). Royal Astronomical Society of Canada.

Raymo, Chet. 365 *Starry Nights.* Prentice Hall, 1982.

Tirion, W. *Cambridge Star Atlas 2000.0.* Cambridge University Press, 1991.

MONTHLY PUBLICATIONS

Astronomy magazine. P.O. Box 1612, Waukesha, Wisconsin 53187.

Sky & Telescope magazine. P.O. Box 9111, Belmont, Massachusetts 02178

Sky Calendar. Abrams Planetarium, Michigan State University, East Lansing, Michigan 48824

Appendix 2: Resource List and Dealers of Astronomy Equipment

Jay Anderson Prairie Weather Center, 900-266 Graham Avenue, Winnipeg, Manitoba R3C 3V4, Canada. Solar and lunar eclipse maps.

Arbor Scientific, P.O. Box 2750, Ann Arbor, Michigan 48106–2750, (800) 367–6695. Science equipment, magnets, and iron filings. A great resource for the teacher.

Astronomical Society of the Pacific, 390 Ashton Avenue, San Francisco, California 94112, (415) 337–2126. Internationally renowned organization dedicated to astronomical education.

Carolina Biological Supply Company, Burlington, North Carolina 27215, (800) 334–5551. Resource materials for astronomy, physics, and a host of other sciences. Good source of magnets, iron filings, and diffraction gratings.

Celestron International, 2835 Columbia Street, Torrance, California 90503, (800) 421–1526. Telescopes, binoculars, and accessories.

Coulter Optical, Inc., P.O. Box K, Idyllwild, California 92349, (714) 659–4621. Inexpensive Newtonian reflectors.

Eastman Kodak Company, Department 454, Rochester, New York 14650, (800) 242–2424. Request Kodak Publication P150, *Astrophotography Basics,* and Kodak Publication L-5, *Index to Kodak Information.*

Edmund Scientific, 101 East Gloucester Pike, Barrington, New Jersey 08007-1380, (609) 547–3488. Science kits and equipment, telescope equipment, diffraction gratings.

Fred Espenak, NASA Goddard Space Flight Center, Planetary Systems Branch, Code 693, Greenbelt, Maryland 20771, (301) 282–8570. Solar and lunar eclipse maps.

International Dark-Sky Association, 3534 North Stewart, Tucson, Arizona 85716, (602) 325–9346. Light-pollution advocacy group.

Meade Instruments Corporation, 16542 Millikan Avenue, Irvine, California 92714, (714) 756–2291. Telescopes and accessories.

Novak and Company, Box 69, Ladysmith, Wisconsin 54848, (715) 532–5102. Telescope-making components.

Orion Telescope Center, P.O. Box 1158, Santa Cruz, California 95062, (800) 447–1001, or (800) 443–1001 (in California only). Telescopes, binoculars, components, and accessories.

Parks Optical, 270 Easy Street, Simi Valley, California 93065, (805) 522–6722. Telescopes, components, and accessories.

Thousand Oaks Optical, Box 5044–289, Thousand Oaks, California 91359, (805) 491–3642. Solar filters and viewers, including Mylar sun viewers.

Appendix 3: Lunar Phase Dates (1994–2000)

The following table lists the dates of the lunar phases for the years 1994 through 2000. Note that some exact phase dates may differ because they may occur near midnight. The actual times of the phases are omitted, as they depend on one's location on Earth, but can be found by consulting the monthly and annual periodicals listed in Appendix 1.

Key to symbols

* = Months with two full moons (the second is known as "Blue Moon")
LE = Some type of lunar eclipse (see Table 25-1)
SE = Some type of solar eclipse (see Table 34-1)
!! = Two eclipses in one month (either solar, lunar, or both)

	NEW MOON	FIRST QUARTER	FULL MOON	LAST QUARTER
1994				
January	11	19	27	4
February	10	18	25	3
March	12	20	27	4
April	10	18	25 2	
May !!	10 (SE)	18	24-25 (LE)	2
June	9	16	23	30
July	8	15	22	30
August	7	14	21	29
September	5	12	19	27
October	4	11	19	27
November !!	3 (SE)	10	18 (LE)	26
December	2	9	17	25
1995				
January	1, 30	8	16	23
February	—	7	15	22
March	1, 30	9	16	23
April !!	29 (SE)	8	15 (LE)	21
May	29	7	14	21
June	27	6	12	19
July	27	5	12	19
August	25	3	10	17
September	24	2	8	16
October !!	23-24 (SE)	1, 30	8 (LE)	16
November	22	29	7	15
December	21	28	6	15
1996				
January	20	27	5	13
February	18	26	4	12
March	19	26	5	12
April !!	17 (SE)	25	3-4 (LE)	10

	NEW MOON	FIRST QUARTER	FULL MOON	LAST QUARTER
May	17	25	3	10
June	15	24	30	8
July	15	23	30	7
August	14	21	28	6
September	12	20 26-	27 (LE)	4
October !!	12 (SE)	19	26	4
November	10	17	24	3
December	10	17	24	3
1997				
January	8	15	23	1, 31
February	7	14	22	—
March !!	8-9 (SE)	15	23-24 (LE)	2, 31
April	7	14	22	29
May	6	14	22	29
June	5	12	20	27
July	4	12	19	26
August	3	11	18	24
September !!	1 (SE)	9	16 (LE)	23
October	1, 22	9	15	22
November	29	7	14	21
December	29	7	13	21
1998				
January	28	5	12	20
February	26 (SE)	3	11	19
March	27	5	12 (LE)	21
April	26	3	11	19
May	25	3	11	18
June	23	1	9	17
July	23	1, 31	9	16
August !!	21-22 (SE)	30	7 (LE)	14
September	20	28	6 (LE)	12
October	20	28	5	12
November	18	26	4	10
December	18	26	3	10
1999				
January *	17	24	1, 31 (LE)	9
February	16 (SE)	22	—	8
March *	17	24	2, 31	10
April	15	22	30	8
May	15	22	30	8
June	13	20	28	6
July	12	20	28 (LE)	6
August	11 (SE)	18	26	4

	NEW MOON	FIRST QUARTER	FULL MOON	LAST QUARTER
September	9	17	25	2
October	9	17	24	1, 31
November	7	16	23	29
December	7	15	22	29
2000				
January	6	14	20-21 (LE)	28
February	5 (SE)	12	19	26
March	6	13	19	27
April	4	11	18	26
May	3	10	18	26
June	2	8	16	24
July !!	1 (SE), 30-31 (SE)	8	16 (LE)	24
August	29	6	15	22
September	27	5	13	20
October	27	5	13	20
November	25	4	11	18
December	25 (SE)	3	11	17

Appendix 4: Planet Positions (1994–2000)

Finding the four brightest planets is easy, once you know where to look. This table gives the constellations where Venus, Mars, Jupiter, and Saturn will be at the middle of each month. Each constellation name is with its standard three-letter abbreviation. Definitions are given at the end. An *italicized* constellation means the planet will be visible in the early evening sky.

	VENUS	MARS	JUPITER	SATURN
1994				
January	Sgr	Sgr	Lib	Aqr
February	Aqr	Cap	Lib	Aqr
March	*Psc*	Aqr	Lib	Aqr
April	*Ari*	Psc	Lib	Aqr
May	*Tau*	Ari	*Vir*	Aqr
June	*Cnc*	Ari	*Vir*	Aqr
July	*Leo*	Tau	*Vir*	Aqr
August	*Vir*	Tau	*Vir*	*Aqr*
September	*Vir*	Gem	*Lib*	*Aqr*
October	*Lib*	Cnc	*Lib*	*Aqr*
November	Vir	Leo	Lib	*Aqr*
December	Vir	Leo	Sco	*Aqr*

Key to constellation abbreviations

Aqr—Aquarius the Water Bearer
Ari—Aries the Ram
Cnc—Cancer the Crab
Cap—Capricornus the Sea-goat
Gem—Gemini the Twins
Leo—Leo the Lion
Lib—Libra the Balance Scale
Psc—Pisces the Fishes
Sgr—Sagittarius the Archer
Sco—Scorpius the Scorpion
Tau—Taurus the Bull
Vir—Virgo the Maiden

Key to constellation abbreviations

Aqr—Aquarius the Water Bearer

Ari—Aries the Ram

Cnc—Cancer the Crab

Cap—Capricornus the Sea-goat

Gem—Gemini the Twins

Leo—Leo the Lion

Lib—Libra the Balance Scale

Psc—Pisces the Fishes

Sgr—Sagittarius the Archer

Sco—Scorpius the Scorpion

Tau—Taurus the Bull

Vir—Virgo the Maiden

	VENUS	MARS	JUPITER	SATURN
1995				
January	Lib	*Leo*	Sco	*Aqr*
February	Sgr	*Leo*	Sco	*Aqr*
March	Cap	*Leo*	Sco	Aqr
April	Aqr	*Leo*	Sco	Aqr
May	Psc	*Leo*	*Sco*	Psc
June	Tau	*Leo*	*Sco*	Psc
July	Gem	*Vir*	*Sco*	Psc
August	Leo	*Vir*	*Sco*	*Psc*
September	Vir	*Vir*	*Sco*	*Psc*
October	*Vir*	*Lib*	*Sco*	*Aqr*
November	*Sco*	*Sco*	*Sco*	*Aqr*
December	*Sgr*	*Sgr*	Sgr	*Aqr*
1996				
January	*Cap*	Sgr	Sgr	*Aqr*
February	*Psc*	Aqr	Sgr	*Psc*
March	*Ari*	Psc	Sgr	Psc
April	*Tau*	Psc	Sgr	Psc
May	*Tau*	Ari	*Sgr*	Psc
June	Tau	Tau	*Sgr*	Psc
July	Tau	Tau	*Sgr*	Psc
August	Gem	Gem	*Sgr*	*Psc*
September	Cnc	Cnc	*Sgr*	*Psc*
October	Leo	Leo	*Sgr*	*Psc*
November	Vir	Leo	Sgr	*Psc*
December	Lib	Vir	Sgr	*Psc*
1997				
January	Sco	Vir	Sgr	*Psc*
February	Cap	Vir	Cap	*Psc*
March	Aqr	*Vir*	Cap	Psc
April	Psc	*Leo*	Cap	Psc
May	*Tau*	*Leo*	Cap	Psc
June	*Gem*	*Vir*	Cap	Psc
July	*Cnc*	*Vir*	Cap	Psc
August	*Vir*	*Vir*	Cap	Psc
September	*Lib*	*Lib*	*Cap*	*Psc*
October	*Sco*	*Sco*	*Cap*	*Psc*
November	*Sgr*	*Sgr*	*Cap*	*Psc*
December	*Cap*	*Sgr*	*Cap*	*Psc*

	VENUS	MARS	JUPITER	SATURN
1998				
January	Sgr	*Cap*	*Cap*	*Psc*
February	Sgr	*Aqr*	Aqr	*Psc*
March	Cap	*Psc*	Aqr	*Psc*
April	Aqr	Ari	Aqr	Psc
May	Psc	Tau	Aqr	Psc
June	Ari	Tau	Psc	Psc
July	Tau	Gem	Psc	Psc
August	Cnc	Gem	Psc	Ari
September	Leo	Cnc	*Aqr*	*Ari*
October	Vir	Leo	*Aqr*	*Ari*
November	Lib	Leo	*Aqr*	*Ari*
December	Sgr	Vir	*Aqr*	*Ari*
1999				
January	Cap	Vir	*Psc*	*Ari*
February	Psc	Lib	*Psc*	*Ari*
March	Psc	Lib	Psc	*Ari*
April	*Tau*	Lib	Psc	Ari
May	*Gem*	*Vir*	Psc	Ari
June	*Cnc*	*Vir*	Ari	Ari
July	*Leo*	*Lib*	Ari	Ari
August	*Leo*	*Lib*	Ari	Ari
September	Leo	*Sco*	*Ari*	*Ari*
October	Leo	*Sgr*	*Ari*	*Ari*
November	Vir	*Sgr*	*Ari*	*Ari*
December	Lib	*Cap*	*Psc*	*Ari*
2000				
January	Sco	*Aqr*	*Psc*	*Ari*
February	Sgr	Psc	Psc	Ari
March	Aqr	*Psc*	*Ari*	*Ari*
April	Psc	*Ari*	*Ari*	*Ari*
May	Ari	*Tau*	Ari	Ari
June	Tau	Tau	Tau	Tau
July	*Cnc*	Gem	Tau	Tau
August	*Leo*	Cnc	Tau	Tau
September	*Vir*	Leo	Tau	Tau
October	*Lib*	Leo	Tau	Tau
November	*Sgr*	Vir	*Tau*	*Tau*
December	*Cap*	Vir	*Tau*	*Tau*

Key to constellation abbreviations

Aqr—Aquarius the Water Bearer
Ari—Aries the Ram
Cnc—Cancer the Crab
Cap—Capricornus the Sea-goat
Gem—Gemini the Twins
Leo—Leo the Lion
Lib—Libra the Balance Scale
Psc—Pisces the Fishes
Sgr—Sagittarius the Archer
Sco—Scorpius the Scorpion
Tau—Taurus the Bull
Vir—Virgo the Maiden

Appendix 5: Astronomical Scavenger Hunt: Our Results

AUTOMOBILES

Chevrolet *Astro* van (astro, as in space)
Chevrolet *Nova* (an exploding star)
Chevrolet *Vega* (brightest star in the summer constellation Lyra)
Dodge *Aries* (the constellation of the Ram)
Ford *Aerostar* (star)
Ford *Taurus* (the constellation of the Bull)
Ford *Galaxie* (a spelling variation of galaxy)
Mercury (planet; the Mercury automobile used to have models called the *Meteor, Comet,* and *Scorpio*)
Mitsubishi *Eclipse* (either solar or lunar)
Nissan *Pulsar* (remnant of a giant star that exploded)
Pontiac *Astre* (version of "aster," meaning star)
Pontiac *Sunbird* (sun)
Saturn (planet)
Subaru (Japanese for the Pleiades star cluster)
Toyota *Starlet* (star)

MOVIES

(not counting the countless science-fiction films)

Beetlejuice movie and Saturday-morning cartoon (a spelling variation of the bright star Betelgeuse in the winter constellation Orion)

Orion motion pictures (winter constellation)

CANDY

Mars candy bar (planet)
Milky Way candy bar (our galaxy)
Starburst candies (star)

PRODUCTS/FOOD/HOUSEHOLD ITEMS

Comet brand rice
Comet cleanser
Comet brand sugar cones
Croissants (crescent-shaped rolls for moon phase)
La Estrella brand biscuit cookies (Spanish for "star")
Pegasus (flying-horse symbol of Mobil gas, and also an autumn constellation)
Star (symbol for Texaco gasoline: "put a star in your car" used to be their slogan)
Starkist tuna (star)
Sunbeam bread (sun)
Sunkist oranges (sun)
Sunoco gasoline (short for Sun Oil Company)

FLAGS THAT INCLUDE STARS

Australia
Alaska
Brazil
Chile
China
Cuba
Japan
Malaysia
New Guinea
New Zealand
Pakistan
Singapore
Taiwan
Texas
Turkey
United States of America

SONGS

"Twinkle, Twinkle Little Star"
"Blue Moon"
"Aquarius" (autumn constellation)
"Total Eclipse of the Heart"
"Moon River"
"Dark Side of the Moon" (album)
"Venus" (planet)
"Stardust"
"There's a Moon Out Tonight"
"You Are My Shining Star"
"Here Comes the Sun"
"Claire de Lune" (lune as in lunar; moon)
"Moonshadow"

SPORTS TEAMS

Phoenix Suns (basketball)
Minnesota North Stars (hockey)
Houston Astros (baseball)

MISCELLANEOUS

5-*Star* general
Starfish
Starling (a bird whose plumage glistens like stars)
Sunflower
Texas, the Lone *Star* State

Appendix 6: Seasonal Star Atlas

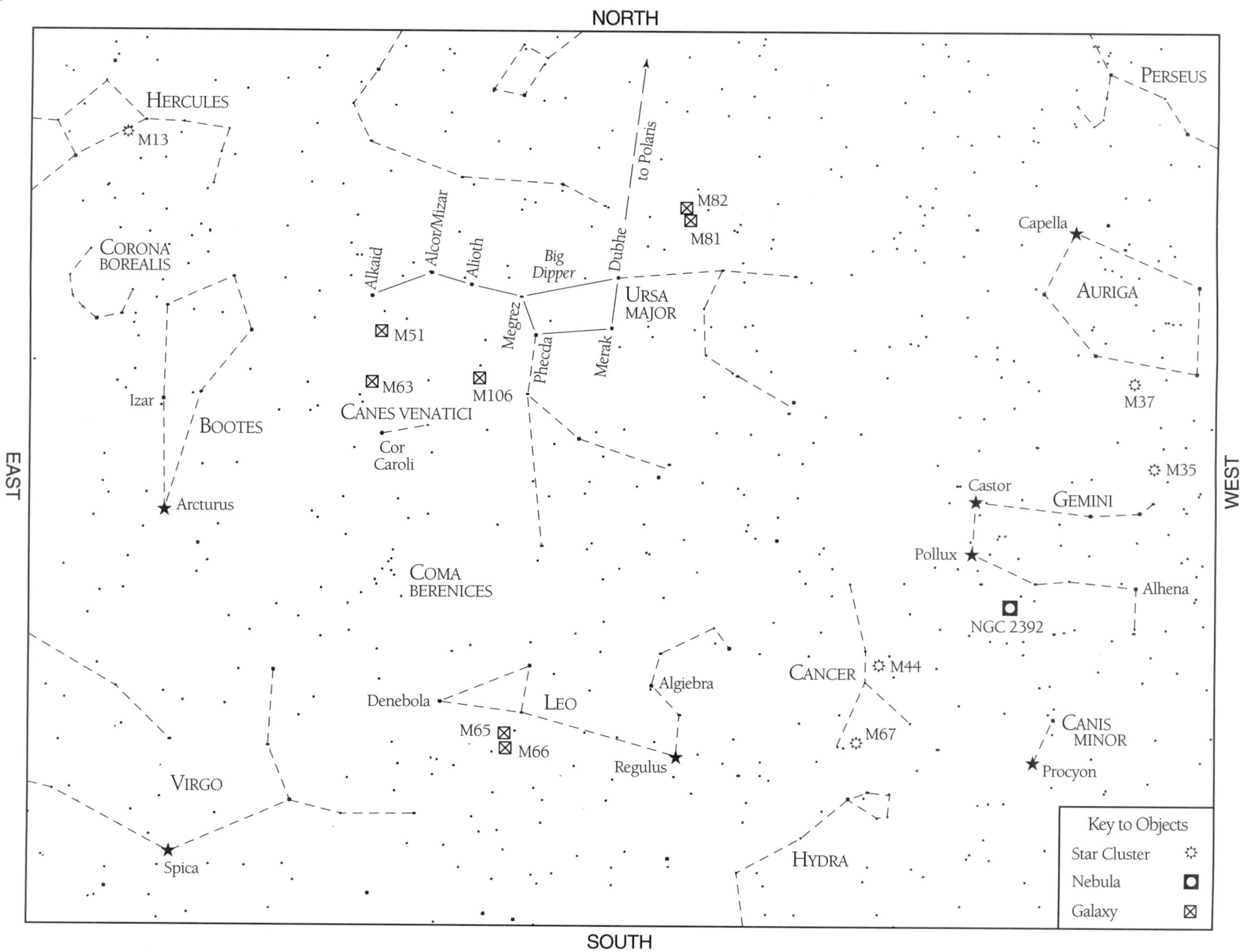

ART 52. Spring sky, showing the positions of some of the deep-sky objects listed in Activities 40 through 43.

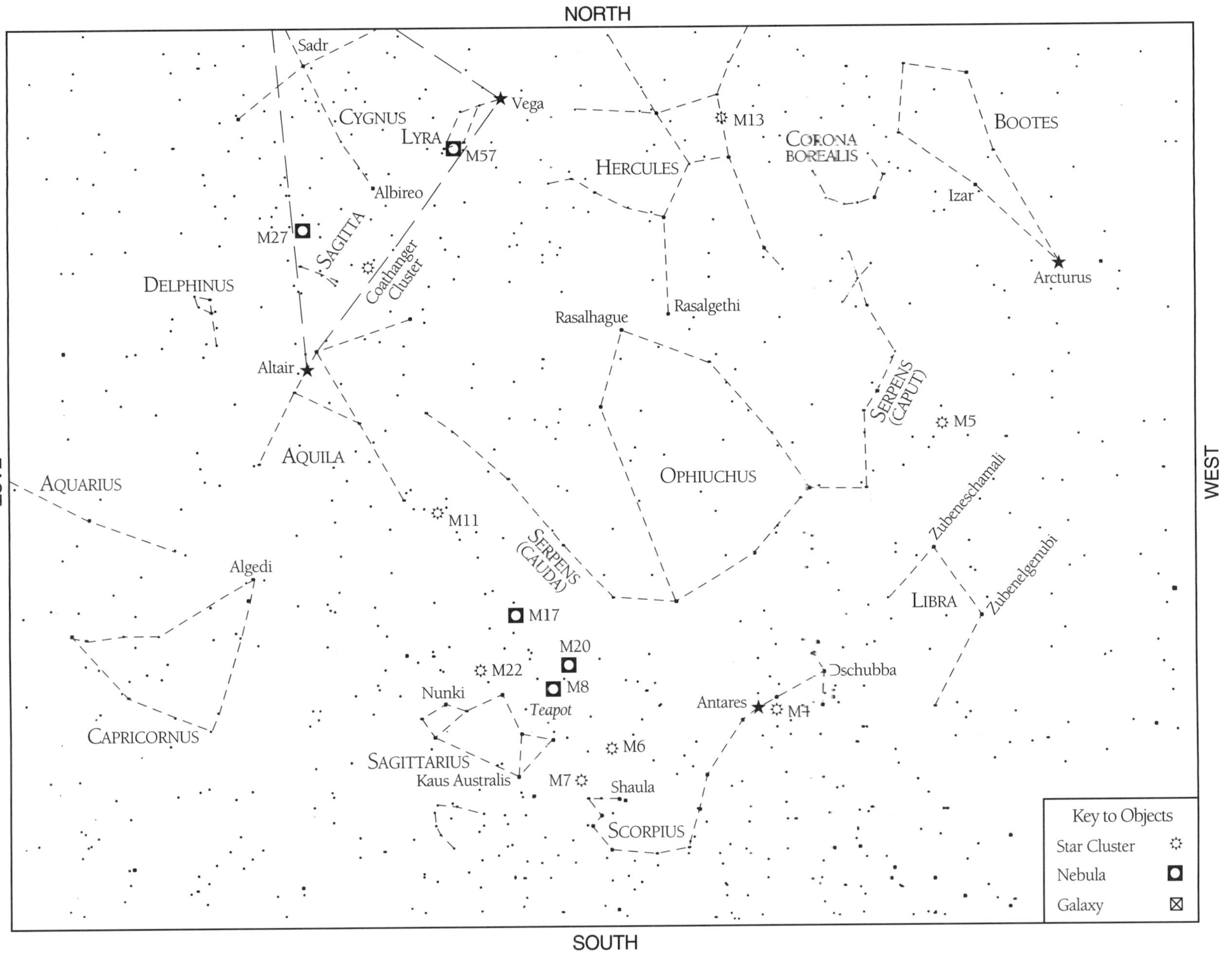

ART 53. Summer sky, showing the positions of some of the deep-sky objects listed in Activities 40 through 43.

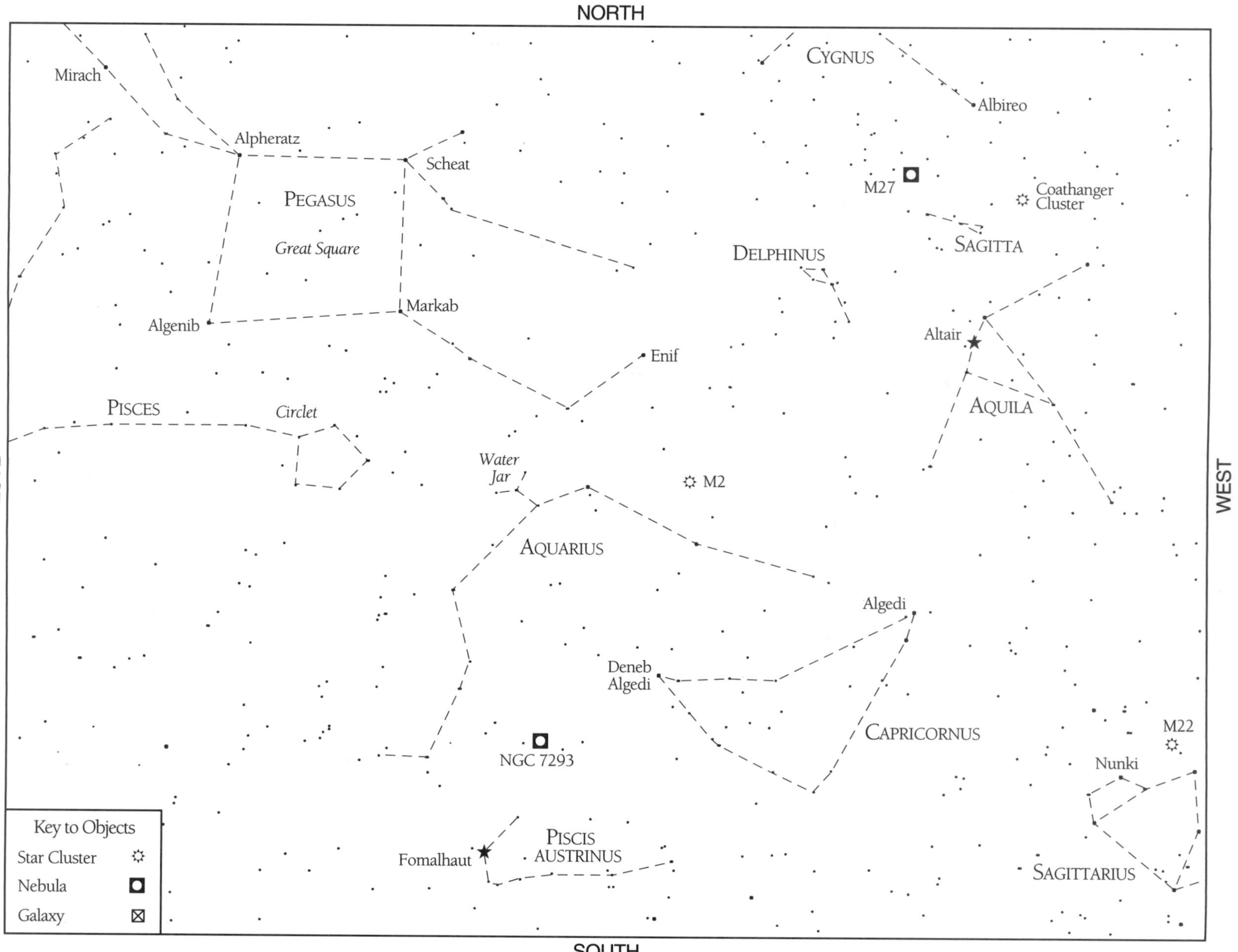

ART 54. Southern autumn sky, showing the positions of some of the deep-sky objects listed in Activities 40 through 43. Note the overlap between this and the next chart.

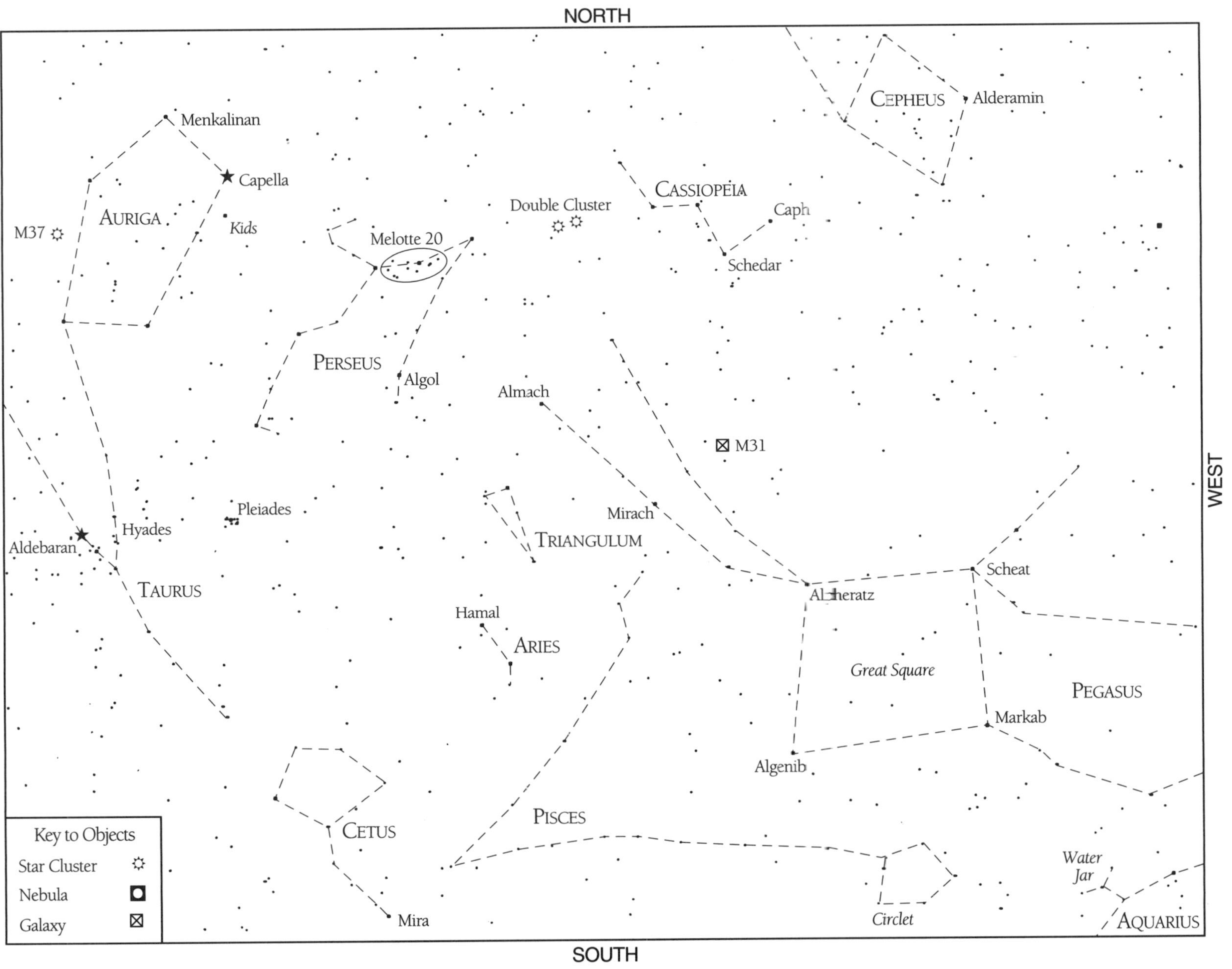

ART 55. Northern autumn sky, showing the positions of some of the deep-sky objects listed in Activities 40 through 43. Note the overlap between this and the previous chart.

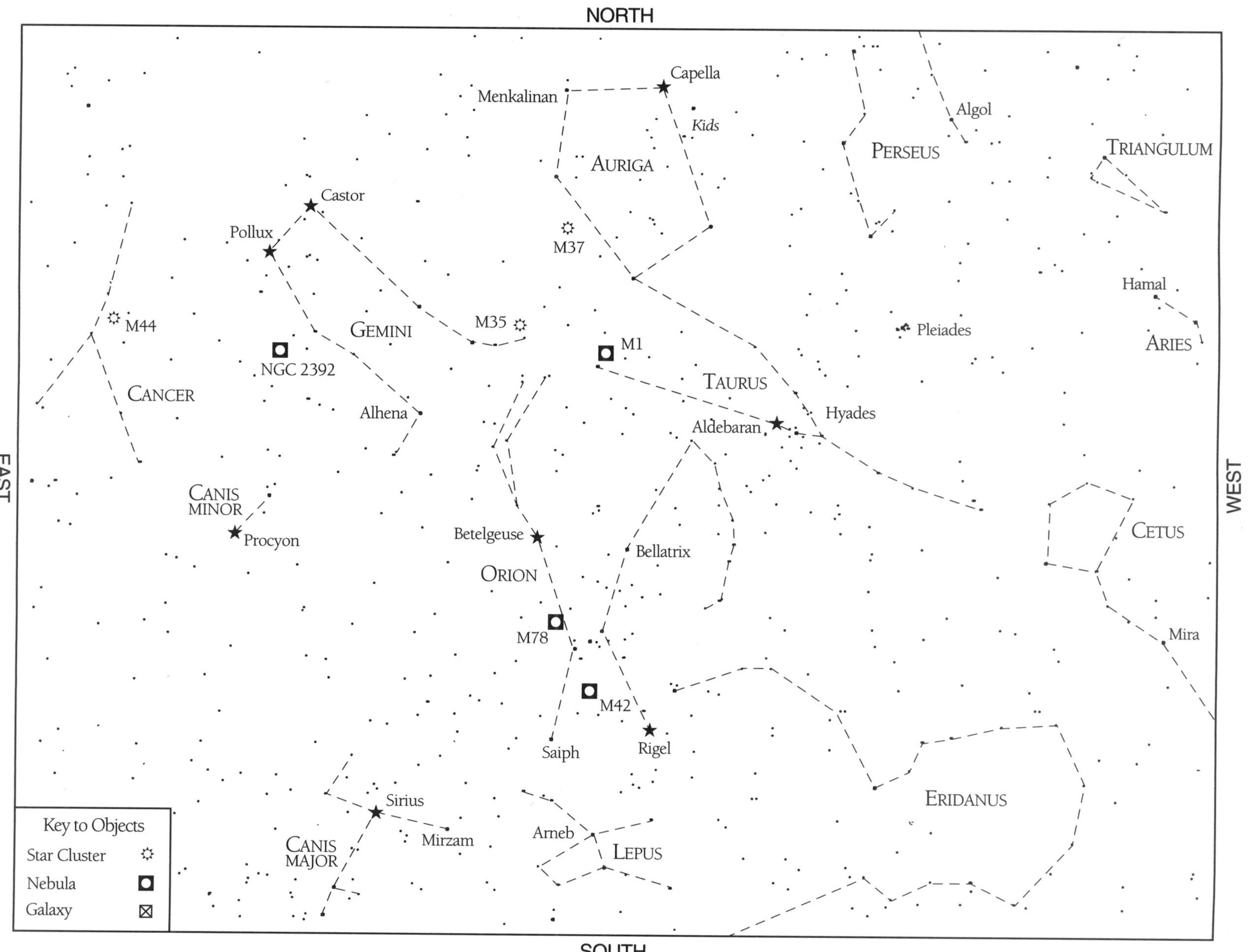

ART 56. Winter sky, showing the positions of some of the deep-sky objects listed in Activities 40 through 43.

Index

ABOUT THE AUTHORS

PHILIP HARRINGTON is the author of *Touring the Universe through Binoculars* and *Star Ware* (John Wiley, 1990 and 1994) as well as numerous articles for *Sky & Telescope* and *Astronomy* magazines. He holds degrees in science education and mechanical engineering. By day he works as an engineer at Brookhaven National Laboratory in Upton, New York. By night he teaches courses in astronomy at the Vanderbilt Museum Planetarium and at Hofstra University and is a frequent lecturer at astronomy clubs and conventions nationwide. He lives in Smithtown, New York, with his wife, Wendy, and daughter, Helen.

EDWARD PASCUZZI, a high school physics and earth science teacher, holds a Bachelor's degree in theoretical astrophysics and a Master's in physics education. For the past decade he has also lectured and taught astronomy at New York's Vanderbilt Museum Planetarium. He frequently contributes material and book reviews to *The Physics Teacher* and *The Science Teacher* and is an avid astrophotographer and award-winning nature photographer. He lives in Greenlawn, New York.